Building the *St. Helena II*

Building the *St. Helena II*

Rebirth of a Nineteenth-Century Canal Boat

Carroll M. Gantz

The Kent State University Press

KENT, OHIO

Library of Congress Catalog Card Number 2011047795
ISBN 978-1-60635-122-2
Manufactured in the United States of America

LIBRARY OF CONGRESS CATALOGING-IN-PUBLICATION DATA
Gantz, Carroll.
The resurrection of the St. Helena : nineteenth-century Ohio canal boat / Carroll Gantz.
p. cm.
Includes bibliographical references and index.
ISBN 978-1-60635-122-2 (hbk. : alk. paper) ∞
1. Canals—Ohio—History. 2. St. Helena II (Canal boat : 1970) 3. Canalboats—Ohio.
I. Title. II. Title: Nineteenth-century Ohio canal boat.
TC624.O3G36 2012
386'.4709771—dc23
2011047795

16 15 14 13 12 5 4 3 2 1

This book is dedicated to all the good citizens of Stark County, Ohio, who, from 1964 to 1970, in the village of Canal Fulton on the Ohio & Erie Canal, diligently and enthusiastically participated in the conception, promotion, design, construction, restoration, and operation of the *St. Helena II,* a hand-built replica of a historically authentic nineteenth-century canal boat, the first of its kind in the nation since 1925. Those volunteers that I can recall are mentioned by name, but there are hundreds of others who participated in this community project without recognition. Their combined efforts created a remarkably popular tourist attraction that benefited the local economy and became a Canal Fulton legacy and, much more significantly, inspired state and national interest in the preservation and recreational reuse of the country's historic canal lands, which were on the verge of being lost forever.

Contents

Preface

In *Doctor Faustus,* dramatist Christopher Marlowe (1564–1593) described the beautiful Helen of Troy from Homer's *Iliad* as "the face that launched a thousand ships." It could be said that another lady, *St. Helena II*—a faithful replica of an actual nineteenth-century Ohio canal boat created in Canal Fulton, Stark County, Ohio, in 1970—was "the ship that launched a thousand places."

These places, many described in this book, were those throughout the country, since 1970 that were inspired to restore and re-create canal days with authentic and operational replicas of canal boats, with the restoration of canal locks, canal towpaths, canal villages, and the canals themselves. Ranging from small, rural communities to major metropolitan parks, they all had something in common: a historical and physical location on the once vast interstate system of canals that opened the interior of the country to commerce, industry, and immigration in the early nineteenth century.

In most urban locations, by 1960 the canal system had been obliterated by 135 years of suburban sprawl, highway and industrial development, unrestrained vegetation, and loss of water supply. But here and there, usually in more remote, rural spots, tiny segments of the watered canal remained, with perhaps a crumbling lock or surviving towpath, which over the years had became a small park for picnicking, hiking, fishing, or canoeing.

The small village of Canal Fulton in Stark County, Ohio, with a mere two thousand residents, was one of these places. "Progress" had passed it by, as it had many other similar rural locations along the old Ohio & Erie Canal. No modern development, no new industry, no supermarkets, no McDonalds—just seventy-year-old Victorian buildings, churches, and homes, and a canal landscape that had remained virtually unchanged since the 1830s. Many residents found this a quaint setting, but nevertheless, the village struggled with a diminishing population and financial resources.

Gradually, inspired by the vision of then local Navarre village solicitor Ralph Regula and *Canton Repository* reporter Al Simpson, the village fathers realized that the canal segment and lock that remained were not merely historical artifacts reminding them of their community's heritage, but potential resources that could become an opportunity to develop, enhance, and create an interactive, historical attraction that could benefit local businesses, attract visitors, and raise community pride. The community built the *St. Helena II,* a living resurrection of the original *St. Helena,* a nineteenth-century canal boat. Without any significant village financial resources, residents had to rely on willing and talented volunteers, charitable donations, hard work, and pure ingenuity. They accomplished their task despite the 1960s being a very tumultuous and tragic decade. This book, in chapters 4 through 8, describes exactly how they did it. Beginning in chapter 5, the narrative shifts to first person, since I became personally involved as a volunteer and shared in both their challenges and their accomplishments.

The achievement was not only financially successful but a booster of civic pride for Canal Fulton and Stark County. More significant, the accomplishment had an impact that spread far. Over the next forty years, the strategy and template for success created by Canal Fulton was communicated through local and national media and, community by community, inspired many other small villages to build similar replica boats and restore their local portion of historic canals. On a larger scale, regional planners and national parks joined with such local efforts in the restoration of hundreds of miles of old canal lands for public recreational and educational reuse, through networking and collaboration of local, regional, and national organizations. Today, millions of Americans enjoy the "thousands of places" launched by the *St. Helena II* in 1970. Chapters 9 and 10 describe just a few.

Canal restoration recaptured a national heritage almost lost forever. To gain a perspective of the importance of canals to the formation of this nation, one must go back to only two years after the fledgling United States gained formal independence from England, as it addressed challenging problems of transportation that hindered travel, commerce, and industry, and prevented the settlement of the vast wilderness frontier beyond the Allegheny Mountains. Canals became a successful solution to all these problems. They were the country's first interstate highway system and its first major national accomplishment after independence. Chapters 1 and 2 describe how this happened, and chapter 3 details how canals disappeared as the railroads took over their transportation function.

Acknowledgments

I am indebted to all those who participated in the resurrection of the *St. Helena* canal boat and to some of them or their families who assisted me in writing this book. Among the latter are fellow volunteer John C. Harriman, son of Ed Harriman, for outstanding photos from the Edward W. Harriman Family Collection and for information enabling me to contact other participant families. John's published book and digital presentation were invaluable references.

Another fellow volunteer of *St. Helena II* construction, Jim Guest, currently a trustee of the Canal Fulton Heritage Society who also serves as a costumed docent on the *St. Helena III,* provided me with information regarding other participants and photos of the construction era. He and his wife, Barbara, graciously hosted my visit to Canal Fulton in August 2010, and Jim gave me an extensive tour of the canal restoration from Akron to Cleveland.

Al Simpson, the journalist who inspired the *St. Helena II,* wrote a book, *Along the Towpath,* which provided me with valuable information. Victoria Neidert-Hammer, Pete Neidert's granddaughter, provided critical information and a photo of Pete's pioneer pontoon boat that initiated Canal Fulton canal boat rides. Pete's son, Donald Neidert, also provided helpful information. Mule driver Glenn Oser's daughter, Bonnie Finefrock, provided information regarding the early mule teams that powered the

St. Helena II. Her husband, Terry Finefrock, became the second captain of the *St. Helena II.*

Scott McLaughlin generously granted me permission to use photos from McLaughlin family collections, and his mother, Ann McLaughlin Henman, provided valuable family information. Ed Shuman, one of the subsequent captains of the *St. Helena II,* provided valuable historical images and information about the McLaughlin drydock.

I am grateful to a number of historical organizations for their generous help. Deborah Lang, executive director of the New Hope (Pennsylvania) Historical Society, put me in touch with Susan Taylor, of the Friends of the Delaware Canal, who provided information on the start of postwar canal barge rides in New Hope. She also put me in contact with George Schweickhardt, head of New Hope canal boat operations between 1977 and 1997, who provided additional information. Ahna Wilson, historian of the C&O Canal National Historical Park, in Washington, D.C., and Potomac, Maryland, provided me with valuable information about all five boats that operated out of Georgetown and Great Falls Tavern.

John Hatfield, president of the Canal Fulton Heritage Society, and Robert Hodges, administrative assistant, graciously researched photo archives and gave me permission to use historical photos. Mary Ann Higgins, Canalway programs director of the City of Canal Fulton Canalway Center, generously directed me to various people with knowledge of local history. A number of historical images of canal boats are from Clyde Gainey's Old Canal Days Museum, operated by the Canal Fulton Heritage Society. Karl Ash, librarian of the William McKinley Presidential Library and Museum, provided information and offered photos. *Canton Repository* executive editor Jeff Gauger and staff member Linda Patten generously gave me permission to use a photo from their files. Richard Sicha, principal planner of the Cuyahoga County Planning Commission, gave information about James Dillow Robinson. Todd A. Clark, historic programmer for the Stark County Park District, provided specifics regarding the history of Lock 4 Park.

My colleagues at the American Canal Society (ACS), including fellow director Terry Woods and the late Thomas Hahn and William Shank, were invaluable sources, through their many articles and books about canals, as were the thirty-eight years of issues of the ACS newsletter, *American Canals.* The books and articles of many such authors provide a rich and detailed history of the canals far more complete than the brief summaries in my book. In a similar manner, *Canal Fever,* by Lynn Metzger and Peg Bobel, describes eloquently and thoroughly the background, evolution, and results of the renovation and reclamation of the Ohio & Erie Canal since the late 1960s, much more completely than the summary in this book allows.

Special acknowledgment goes to Terry Woods, of Canton, Ohio, canal historian, educator, author, and past president of both the Canal Society of Ohio and the American Canal Society; and to Peg Bobel, of Akron, Ohio, canal historian, author, conservationist, cofounder of the Cuyahoga Valley Trails Council, and past executive director of the Cuyahoga Valley Association. Terry and Peg generously reviewed my manuscript, providing invaluable information and insight.

I am also grateful to Kent State University Press for support, advice, guidance, and expert copyediting by Erin Holman.

Without the assistance of so many, this book could not have been written. I am deeply grateful to all.

CHAPTER 1

American Canals

After 1825, before railroads rendered them functionally superfluous, canals were the primary and most efficient interstate transportation system of the new United States for about thirty years. At their peak, there were 117 canals (53 major ones) in nineteen states (there were only twenty-four states in 1825), ranging in length from a half of a mile to 363 miles (seven were 200 miles or longer), for a total of 4,700 miles. The canals provided passengers, settlers, and immigrants with overland travel more comfortable and less expensive than dusty stagecoaches or wagons on unpaved roads and, more important, provided inexpensive transport for raw materials and commercial cargo between deepwater ports and the vast wilderness territories of farms and isolated communities. The fundamental legacy of the canal systems is that they bound the early United States together in industry and commerce, fostered massive immigration into the expanding frontier, and enabled the country to compete with Europe in the Industrial Revolution.

Canals had been built since antiquity. China's Grand Canal and Nebuchadnezzar's Royal Canal of Babylon were constructed in 500 B.C. Leonardo da Vinci designed canals and canal locks in Milan in the fifteenth century. Canals had been used in France since the seventeenth century, and England built canals linking saltworks to Liverpool, with the River Mersey, between 1713 and 1741. The Bridgewater (Worsley) Canal, which transported coal from Lancashire to Manchester, was built in 1761. By 1792 there were thirty canal schemes in England competing for capital, since canals were essential to support the English Industrial Revolution, by then transforming the country economically. Americans were well aware of canal technology and understood its potential for the development of commerce and industry. Transportation in colonial America was essentially limited to seaports, the Great Lakes, and deep-river access by sailing ships. A number of major rivers reached inland to city ports such as New York, Boston, Baltimore, Philadelphia, Pittsburgh, and Savannah, but beyond them lay a vast, expanding frontier with impossible access by sail. As they climbed inland toward their sources, rivers became shallower and narrower, and were blocked by rapids, preventing access by large ships. *(See map in color insert, middle of book.)*

Canals, beside and watered by the smaller rivers in inland valleys, could tame these rivers by creating level stretches so that large flat-bottomed boats with shallow draft could be pulled inland by horses or mules. Hydraulic locks, huge boxlike structures, could be filled with and emptied of water, permitting boats to be raised or lowered to follow the riverbank topography. In this way, boats could reach far into the American wilderness. The problem was the enormous cost of construction.

Former colonies, now states responsible for their own commercial development, looked to canals to improve river navigation, then the only means of transporting heavy cargo and people inland. In 1792, the State of Pennsylvania began construction of the one-mile Conewago Canal to bypass treacherous Conewago Falls on the Susquehanna River at York Haven, Pennsylvania, eleven miles south of Harrisburg; it opened in 1797. Also in 1792, Pennsylvania initiated a navigable waterway to connect the Susquehanna and Schuylkill Rivers, but in 1794, after fifteen miles had been completed, financing was exhausted, and it would languish until work resumed in 1815, after the war of 1812.

In 1793, the Western Inland Lock Navigation Company built the one-mile long Little Falls Canal with five locks to pass the forty-foot fall on the Mohawk River in New York. It was completed in 1795. Also in 1793, a close friend of George Washington, Revolutionary War major general William Moultrie (1730–1805), governor of South Carolina and president of the Santee Canal Company, began construction of the twenty-two-mile Santee Canal, which enabled crops from the uplands to reach the port of Charleston directly, rather than by wagon or along the unpredictable coastal waters of the Atlantic. It was the first summit-level canal, that is, one for which lakes, rather than rivers, provided water for operation. The ten-lock canal opened in 1800 and continued operation until 1850.

After the American Revolution, convinced of the importance of canals to the new United States, George Washington (1732–1799) hoped the Potomac River could be made navigable westward to the Ohio River Valley and thus open the frontier. In 1785, he founded the Potowmack Company, which built a canal from Georgetown, near where the new capital city of Washington, D.C., was being built, that skirted the rapids of the Potomac River in Virginia and allowed boats to reach upstream to Harpers Ferry. The famous Watergate complex in Washington was named after a series of locks with a water gate opening into the Potomac River. The canal (westward from Georgetown) included Little Falls (two miles, three locks, already built in the 1770s by canal promoter John Ballendine), Great Falls (nine miles, five locks; the most difficult), Seneca Falls, and House Falls, the latter two near Harpers Ferry and the easiest to construct. The Potowmack Canal became operational in 1802 but was engineered poorly, thus making it difficult to navigate upstream against a swift current. The company went bankrupt, by 1823 the canal was considered a failure, and in 1824 it was ceded to the Chesapeake & Ohio Company, which would rebuild it starting in 1828.

The Middlesex Canal in Massachusetts, chartered in 1793 by Governor John Hancock (1737–1793) and built between 1795 and 1803, connected the Merrimack River near Lowell, an industrial center, with Boston Harbor, a distance of twenty-two miles, with twenty locks. It cut in half the previous route down the Merrimack River to the seaport at Newburyport on the Atlantic, and thence to Boston.

These early efforts were of a local nature only and were discouraged by costs generally too large for private enterprise. Robert Fulton (1765–1815), who would in 1807 devise the first commercially successful steamboat, had studied canal technology in England and as early as 1796 had written a treatise on canal design and construction, containing detailed drawings of vertical lifts, aqueducts, and bridges. He submitted copies to Governor Thomas Mifflin of Pennsylvania (1744–1800), who had appealed to Congress for more affordable transportation systems since 1795.

Other states with transportation problems were already planning similar local canals but were stymied by the immense cost. To reach westward to the frontier beyond the coastal areas and Appalachian Mountains with a connected network, a federal program was needed to finance and encourage development of larger transportation systems. In 1807, the U.S. Senate passed a resolution calling for the Treasury Department to compile a report proposing ways that the federal government could address the transportation problems in the new nation. The task fell to Albert Gallatin (1761–1849), treasury secretary in Jefferson's cabinet, who in 1808 produced a landmark "Report on Roads, Canals, Harbors, and Rivers" that proposed, among other ideas, a series of canals parallel to the Atlantic coast from New York City to South Carolina; a major turnpike from Maine to Georgia; a canal crossing New York state; a series of inland canals heading to Ohio; sand improvements to make rivers, including the Potomac, Susquehanna, James, and Santee, passable to major river navigation.

This was an innovative plan to open the interior of the country to commerce and immigration. The cost of this transportation infrastructure was estimated at $20 million, and when proposed, many viewed it as impractical for the federal government, though over time many of these concepts were realized. The earliest was the National Road from Baltimore, Maryland, started in 1811 and completed to Illinois in 1839. Today this is known as Route 40, but in the 1830s it was merely gravel that sunk into the mud, and was prone to landslides. Other Gallatin proposals that eventually would be built included the Erie Canal in New York and other canals in New Jersey, Pennsylvania, and Ohio; the Atlantic Intracoastal Waterway; and a north-south Route 1 along the Atlantic coast.

The War of 1812 not only delayed action but further dramatized the need for better national transportation systems for military purposes. A cannon worth $400 at the foundry in Philadelphia had cost the government $2,000 when it was delivered overland to the Canadian frontier. A barrel of pork for the troops cost $100 to deliver over muddy roads by wagon. Even worse than the cost was the time required for such transport, which severely limited the promptness of military response to threats. Similar problems affected civilian commerce. The nation had adequate transportation along its coasts to encourage international commerce and, with the recent development and use of steamboats, on major rivers and the Great Lakes to support regional trade. But rapidly expanding inland populations suffered because of the difficulty and cost of transporting crops and goods to and from trading centers in the East and South. Last but not least was the ever-increasing flow of immigrants seeking cheaper land in the interior of the country.

The federal government did not take action until 1816. Faced with a persistent economic depression after the war, Congress passed legislation to strengthen and unify the new nation in agriculture, commerce, and industry, and federal help became available. This legislation rechartered the Bank of the United States to promote a single currency, provided federal credit to fund improvement of the nation's roads and canals to expand agricultural markets, and funded manufacturing societies to organize new businesses to reduce reliance on European imports.

States were quick to respond to these national priorities; many had already seen the need for canals, had surveyed potential routes, and were ready to provide their own financing and engineering. Most were financed with state bonds, often bought by English investors but also by Americans.

A canal in New York from the Hudson River to Lake Ontario had been proposed as early as 1768, but surveys did not begin until 1808, by James L. Geddes (1763–1838). In April 1817, a report by canal commissioners, appointed by New York state legislators and headed by DeWitt Clinton (1769–1828), completed a study of the proposed route for a canal through the state from Albany on the Hudson River to Buffalo on Lake Erie, a distance of 363 miles. The Erie Canal would be the first major canal in the United States, and also the longest.

Clinton was elected governor of New York on July 1, 1817, and work on the canal began that July 4 in Rome, New York, in the middle of the state. Construction proceeded east and west at the same time. The canal was forty feet wide at the waterline, twenty-eight feet at its bottom, and four feet deep. A ten-foot-wide towpath was constructed along the bank of the canal to accommodate the horses, mules, or oxen that pulled the boats. Eighty-three locks and eighteen aqueducts were required to cross ravines and rivers and scale 568 feet from the Hudson River to Lake Erie. Eight years later, on October 26, 1825, Governor Clinton presided over the opening celebrations and joined the first through passage from Buffalo to Albany with other dignitaries on the *Seneca Chief.*

Coincidentally, just one month before, on September 27, in England, George Stephenson (1781–1848), at the controls of his locomotive, the *Locomotion,* had opened the world's first public railroad, the Stockton and Darlington, to pull thirty-six freight wagons of coal and flour nine miles at fifteen miles per hour.

In its first year of operation, nineteen thousand boats passed through the Erie Canal, carrying freight such as salt from the saltworks in Onandaga County, as well as over forty thousand immigrants. A ton of freight now cost only $10 to ship, compared with $100 by land. The canal attracted businesses and small communities, provided access to Ohio and the Northwest Territories, and made New York City the greatest port in America. In its first eight years, the canal paid for its initial investment and maintenance in tolls. Land values increased by $100 million. Within twelve years, the canal had recouped the cost of its original construction.

These obvious benefits were recognized and publicized around the nation, and "canal fever," the rush to build canals, spread to almost every state. In the mid 1820s, the forty-five-mile Blackstone Canal was built, connecting Providence, Rhode Island, with Worcester, Massachusetts, along the Blackstone River. It was completed in 1828 and required forty-nine locks for climbing the 438-foot difference in elevation.

In 1824, the Potowmack Company in Washington, D.C., ceded its holdings to the Chesapeake and Ohio Company, which had the intention of reaching the Ohio River at Pittsburgh with a canal. Benjamin Wright (1770–1842), formerly chief engineer of the Erie Canal, was named chief engineer of the new effort, and construction began July 4, 1828, with groundbreaking by president John Quincy Adams. Due to money and labor problems, and the difficult mountain terrain, the canal never reached its destination. By 1850, it spanned only 184 miles, from Georgetown to Cumberland, Maryland, along the Potomac River.

Also in 1824, the New Jersey legislature chartered the Morris Canal and Banking Company to construct a canal connecting Phillipsburg on the Delaware River, eastward through New Jersey to the Hudson River at Jersey City, a distance of 107 miles. The canal had to rise 760 feet, then descend 914 feet; this was accomplished with twenty-three locks and twenty-three inclined planes, where open cars carrying canal boats were lifted by waterwheel-powered winches up and down hills. It was completed to Newark by 1831, and to Jersey City by 1836. By then, canal boats could cross the Delaware River in Phillipsburg via a cable ferry and link up with Pennsylvania canals at Easton on the west side. Another New Jersey canal, the Delaware & Raritan, running forty-four miles from Bordertown on the Delaware River to New Brunswick on the Raritan, was completed in 1834, and a feeder canal twenty-two miles north along the Delaware River's east bank, from Bordertown to Bull's Island near Frenchtown, extended the length to sixty-six miles. In the 1860s, it was used to transport coal from Pennsylvania to New York City.

Pennsylvania's Schuylkill Navigation Company restarted work in 1815 and by 1827 had completed its waterway of 108 miles, combining 46 miles of river slackwater pools with 62 miles of separate canals linking Port Carbon in the coal region to Philadelphia. At Reading, the Schuylkill River connected to the Union Canal, which went west to the Susquehanna River at Middletown, 81 miles total, and was completed in 1828.

In 1825, the same year the Erie Canal was completed, Ohio began construction of an 800-mile canal system, which was completed in 1845. This project, central to our Ohio canal boat story, is detailed in chapter 2.

Also in 1825, the Delaware & Hudson Canal Company began building a 109-mile canal—from the Hudson River near Kingston in New York, to Port Jervis on the Delaware River, then north to cross the Delaware into Pennsylvania at Lackawaxen and thence to Honesdale, Pennsylvania. It opened in 1828 to transport anthracite coal to New York City. In 1829, the company purchased the first steam locomotive to be operated in the United States, the *Stourbridge Lion,* built in England the previous year. In Honesdale, it began operation to carry coal to the canal.

Coal was the primary cargo of Pennsylvania's 48-mile Lehigh Canal, from Mauch Chunk (now Jim Thorpe) south to Easton, where it connected to the 60-mile Delaware Division Canal from Easton to Bristol. Both were completed in 1832 and brought coal the 108 miles from Mauch Chunk to Philadelphia.

In 1826, the Pennsylvania Main Line of Public Works system (known as the "Main Line") was authorized to span Pennsylvania from east to west, roughly 315 miles, over the Allegheny Mountains. Yes, over the mountains! The system was divided into geographical divisions for construction. The Western Division Canal began in Pittsburgh, on the Ohio River. It proceeded eastward up the Allegheny River 104 miles to Johnstown and became operational that far in 1831.

There, to cross the Alleghenies, a thirty-seven-mile Portage Railroad from Johnstown was constructed with an ingenious mechanical system of ten inclined planes, each climbing and descending peaks of several hundred vertical feet. Boats were loaded on flat railway cars. At either end of the plane, cars were attached to continuous hemp cables, which had separate rail tracks for each direction. Cars were then raised and lowered on opposite sides of the plane by a stationary steam engine at the peak. Between the planes, horses and, later, locomotives pulled boats on flat railway cars to the next plane.

The Portage railroad, opened in 1834, connected in Hollidaysburg on the east side of the Alleghenies to the Juniata Division Canal, which had opened in 1832 and spanned 127 miles down the Juniata River to meet the Susquehanna Division Canal. It ran 41 miles south along the west bank of the Susquehanna River. There, boats were towed across the river alongside the Clark's Ferry Bridge, built with a floating towpath, to the east side of the river at Northumberland, where it met the Eastern Division of the canal, operational in 1833, which traveled 46 miles south to Columbia. From Columbia to Philadelphia, a distance of about 75 miles, the federal government constructed a toll-financed railroad, one of the first in the nation, which began operations in 1834. Initially, horses pulled sectional canal boats on flat railcars, but they progressively shared the road with steam locomotives until 1844, when horses were banned.

The Cumberland & Oxford Canal, constructed in Maine between 1828 and 1830, was started just several weeks before the other C & O canal, the Chesapeake & Ohio. The Maine canal linked the port and industries of Portland with lumber and wood products to Sebago Lake, a distance of twenty miles with twenty-seven locks to raise canal boats 270 feet above sea level. A twenty-eight lock allowed boats to rise up to Long Lake and enabled them to reach fifty miles inland. The sixty-four-foot-long boats were equipped with sails to cross the lakes.

In western Pennsylvania, the Beaver and Erie Divisions connecting Pittsburgh on the Ohio River to Lake Erie, about 136 miles, were begun in 1831 but not completed until 1844. When all the Pennsylvania divisions and extensions were added together, they totaled over 1,100 miles.

Indiana joined the race with the Wabash & Erie Canal, started in 1832 and completed in 1853, linking the Great Lakes to the Ohio River. At 460 miles, it was the longest in North America, though it was actually a combination of four canals. What later became the Miami & Erie Canal was begun in 1853 from Toledo, Ohio, on Lake Erie to Junction, Ohio, where it connected to the Wabash & Erie Canal, which went from there to Terre Haute, Indiana. At Terre Haute it completed the Cross-Cut Canal that ran to Worthington, Indiana, and then the Central Canal to Evansville, Indiana, on the Ohio River. During the construction, between 1836 and 1847 the Whitewater Canal in Indiana was built, spanning 76 miles, stretching from Lawrenceburg to Hagerstown, Indiana, on the Ohio River. It saved farmers a several-day overland journey to Cincinnati, but after only a few months of service, a section on the state line, in Harrison, Ohio, washed out and was never rebuilt.

The Illinois & Michigan (I & M) Canal was completed in 1848 before railroads were laid in the area, running ninety-six miles from the Bridgeport neighborhood of Chicago on the Chicago River, to LaSalle-Peru on the Illinois River. The canal was sixty feet wide and six feet deep with seventeen locks and four aqueducts. It operated until 1933, but traffic was significantly reduced after the 1853 opening of the Chicago, Rock Island & Pacific Railroad, which ran parallel to the canal.

In 1855, the Black River Canal met the Erie Canal at Rome, New York, and ran north to Carthage, eighty-five miles, including forty miles of river navigation. It required 109 locks and remained in use until 1920.

In 1887, the U.S. Congress authorized the seventy-five-mile Illinois & Mississippi Canal (the I & M Canal is also called the "Hennepin Canal") to connect the Illinois River near Hennepin west to Rock Island on the Mississippi, after farmers on the upper Mississippi lobbied for a better route to the Great Lakes via this and its sister, the I & M Canal. Due to the inadequacy of the I & M and rail competition, it was virtually useless when completed in 1907. The federal government abandoned it in 1951, and the State of Illinois acquired it in 1969.

These and numerous other minor canals connected to form a growing national canal network, which enabled states to compete for vast, expanding commercial markets otherwise beyond their reach and to encourage immigration to unpopulated regions.

What was the national significance of canals? Before them, the new United States was nearly a century behind England in the Industrial Revolution. Before railroads replaced it, the forty-seven-hundred-mile interstate canal system connecting the northeast states became an engine of industrialization. This would prove a crucial Union weapon in the Civil War, as the Confederacy remained an agrarian, maritime society with few canals and little of the industrial development that came with them. This same industrialization of canal lands would by the end of the nineteenth century enable the United States to surpass England in industry, productivity, and standard of living. Without canals or their accompanying industrialization, the new country would not have been able to compete internationally, at least for another generation. But it was an enormous effort. The physical challenge and technical difficulty in building these canals is evident.

CHAPTER 2

Ohio Canals

Ohio was one of the earliest states to build canals. This wilderness frontier became the seventeenth state in 1803, but mass immigration began only after the War of 1812.

In 1817, just as the Erie Canal was begun, Ohio started planning a major enterprise that would result in a network of over 813 miles of canals and make it the third most prosperous state in the Union. From 1810 to 1820, the state's population had doubled to half a million settlers, mostly clustered along the Ohio River Valley. Ohio needed to provide access inland for settlers from Lake Erie in the north, Pennsylvania in the east, and the Ohio River in the south. Also, since Ohio was essentially an agricultural state, to access eastern produce markets, it needed connections to Pennsylvania and New York canals, a faster and more practical route than the long trip down the Ohio and Mississippi Rivers to New Orleans, even with steamboats, which by 1811 were already paddling these rivers.

Ohio's canal system was begun during the 1818–22 term of Governor Ethan Allen Brown (1776–1852), elected after a canal-building campaign. With his initiative, in 1822 the Ohio legislature created a canal commission to hire surveyors and engineers to study possible routes from Lake Erie to the Ohio River. The commission hired James L. Geddes, the same man who surveyed the Erie Canal. He surveyed five routes via natural rivers systems, all of which crossed the north-south continental divide and required summit-level water supply from lakes. Two—the first via the Black and Muskingum (Killbuck Branch) Rivers and the second via the Grand and Mahoning Rivers—were totally impractical. Three remained: in the east, the Muskingum (Cuyahoga Branch) and Cuyahoga; in the central region, to connect to Columbus, the state capital since 1816, the Sandusky and Scioto; and in the west, the Great Miami and Maumee. Geddes concluded, however, that there was insufficient water on the divide between the Scioto and Sandusky Rivers, and so this central route was abandoned as well.

On February 4, 1825, the Ohio legislature passed the Canal Act, which authorized the construction of the Ohio & Erie Canal, "to provide for the internal improvement of the State of Ohio by navigable canals."[1] It was a circuitous route; the canal was to start at Portsmouth on the Ohio River, head north along the Scioto River to a point eleven miles south of Columbus, then east through the Licking Valley to Newark, and north up the Muskingum, Tuscarawas, and Cuyahoga Rivers to Cleveland on Lake Erie. The act also approved the Miami Canal from Cincinnati on the Ohio River to Dayton on the Miami River, with the hope that it could later be extended northward to Toledo as development proceeded on the Ohio & Erie Canal. Lawmakers in Columbus were mollified by the authorization of an additional eleven-mile "feeder" canal connecting the capital to the Ohio & Erie Canal.

The State was authorized to appropriate property for construction where required. Even before work began, real estate values along proposed routes increased significantly as the State acquired property. Many landowners generously donated property to insure that the canal would benefit them. For example, near Portage Summit, Gen. Simon Perkins (1771–1844), president of the Western Reserve Bank, and Paul Williams (d. 1828) surveyed and platted a new town, to be called Akron, and offered canal commissioners land for a turning basin and a right-of-way directly through the town.

THE OHIO CANALS
(public and private)

Canal	Termini	Construction Begun	Construction Completed	Approximate Cost	Number of Locks*	Number of Miles**	Date Abandoned
OHIO & ERIE	Cleveland - Portsmouth	1825	1834	$4,695,000	152	333	1913
Hocking Valley	Carroll - Athens	1826	1843	975,000	26	56	1894
Sandy and Beaver	Bolivar - Ohio River	1834	1848	1,850,000	90	73	1884
Pennsylvania and Ohio	Akron - Penna. Line	1835	1841	1,300,000	57	86	1877
Muskingum Improvement	Dresden - Marietta	1836	1841	1,627,000	12	91	Open
Walhonding	Brinkhaven - Coshocton	1836	1841	607,000	11	25	1913
MIAMI & ERIE	Toledo - Cincinnati	1825	1845	6,762,000	105	266	1909
Warren County	Middletown - Lebanon	1834	1840	228,000	6	20	1848
Wabash and Erie	Indiana Line - Junction	1837	1843	?	?	18	1854
Cincinnati and Whitewater	Harrison - Cincinnati	1839	1843	800,000	7	25	1863
Milan	Huron River - Milan	1833	1839	23,390	2	3	1865

*Including feeder locks (guard locks not tabulated) **Including navigable feeders

Some of the statistics are the result of the best determination that could be made on the basis of the material available, which contains discrepancies.

MAJOR RESERVOIRS
(all are now Ohio state parks)

Reservoir	Location	Construction Dates	Approximate Cost	Acreage
OHIO AND ERIE				
Licking Summit (Buckeye Lake)	Licking-Fairfield-Perry Counties	1826-32; enlarged 1836	$200,000	originally 4200 acres; enlarged by 500 acres
Portage Summit (Portage Lakes)	Summit County	damming, which enlarged natural chain of small lakes, and Tuscarawas Feeder, completed 1840	80,000	1979 acres
MIAMI AND ERIE				
Mercer County (Grand Lake - St. Marys)	Mercer-Auglaize Counties	1837-41	528,222	originally 17,603 acres; reduced to 13,440 acres 1856
Lewistown (Indian Lake)	Logan County	1851-52; enlarged 1856-60	341,126	originally 1000 acres including pond "Indian Lake;" enlarged to 6134 acres
Loramie Summit (Lake Loramie)	Shelby-Auglaize Counties	completed 1844	22,000	1800 acres

In addition, smaller reservoirs were located on both state and private canals.

On July 4, 1825, at Licking Summit, near Newark, Ohio, Governor DeWitt Clinton of New York, the champion of the Erie Canal, and Ohio governor Jeremiah Morrow (1771–1852) led a groundbreaking ceremony for the Ohio & Erie Canal, three months before the completion of the Erie Canal. Why was Clinton there? Because New Yorkers saw the Ohio Canal as an extension of the Erie Canal, providing a thousand-mile waterway from the Hudson River to Buffalo on Lake Erie, where lake steamers could reach Cleveland, and thence, by the Ohio & Erie Canal, to the Ohio River and south. Many New York financiers had bought Ohio bonds to construct the canal and had a vested interest in its success.

Less than three weeks later, on July 21, the two governors participated in similar groundbreaking ceremonies in Middletown, Ohio, for a twenty-five-mile Miami Canal south to Cincinnati. The latter would be completed in 1828 and by 1830 would extend northward to Dayton, another seventeen miles.

Work began in 1825 on the Ohio & Erie Canal from Portage Summit near Akron to Cleveland on Lake Erie, which at that time had only about two hundred residents. This was one of the most critical and difficult parts of the canal. The first thirty-eight miles from Lake Erie to Portage Summit at Akron required forty-four lift locks to raise the canal a total of 395 feet, twenty-one of them in the last two miles before the summit, fifteen in a single mile, and seven in a steep half-mile "staircase" in Akron, known as Cascade Locks. Each lock in the staircase raised the canal ten feet, for a total rise of seventy feet. Canal commissioner Alfred Kelley (1789–1859) supervised construction on horseback. He had been elected to the Ohio House of Representatives in 1814 and would serve until 1857. Kelley had persuaded the state legislature to finance the canal and in 1815 had also been elected the first mayor of Cleaveland (whose spelling changed to "Cleveland" in 1830).

The construction of the Ohio & Erie was an enormous undertaking. The main canal was 308 miles long, with 146 lift locks to transverse up or down the natural terrain. Locks, most constructed of sandstone or limestone masonry with wooden bottom timbers, were ninety feet long, fifteen feet wide, and up to twenty feet deep, limiting the size of the canal boats using them. At each end were two large wooden lock gates, closed in a "V" that pointed upstream, which could be swung open or closed by long balance beams, also called "sweeps." Initially, locktenders who lived in nearby shanties manually operated them, but later boat crews had to do so. A boat traveling upstream entered the lock through the open, lower gates, at the downstream water level. The lower gates were then closed, and wicket gates opened in the closed upper gate, filling the lock and raising the boat to the upstream water level. The upper gates were then opened, and the boat towed or poled through. Boats traveling downstream followed the same process in reverse but were lowered in the lock rather than raised, by opening the wicket in the lower gates. Canals were essentially enormous hydraulic systems of locks powered by gravity.

Locks were the relatively small, highly technical, components of the system, but they were dwarfed by the Herculean task of digging a 333-mile-long ditch (including feeders and sidecuts). The canal prism itself was forty feet wide at water level, twenty-six feet wide at the bottom, and four feet deep. A ten-foot-wide towpath was constructed, usually between the canal and the river, for the tow animals and their drivers, and on the opposite side of the canal, a six-foot-wide berm bank was required. The bottom and sides of the canal were lined, or "puddled," with clay, to waterproof it.

Throughout the state, contracts were offered for construction of half-mile or longer sections of the canal. Many of the contractors had worked on the recently completed Erie Canal and flocked to Ohio for new contracts. Even local farmers could bid to build segments that crossed their property.

Typical subcontracts included "grubbing and clearing," "mucking and ditching," "embankment and excavation," "locks and culverts," "puddling," and "protection," the last meaning protecting the embankments from erosion. Forty-four aqueducts, essentially water-filled bridges, were needed to carry canals across streams or rivers. By November 1825, more than two thousand men were working on the thirty-eight-mile segment from Cleveland to Portage Summit. The canal was constructed entirely by hand labor, with shovels, wooden scoops, wheelbarrows, and mule-drawn carts. Workmen were predominantly native-born Americans but included Irish, German, and other immigrants, and prison inmates were used as well. Average pay for labor was 30 cents per day. Workers lived in clusters of crudely built shacks that later became towns or major cities. Diseases including cholera, smallpox, typhoid, and malaria were rampant in the swampy environment during intense heat or during cold, damp weather, causing many deaths.

The Ohio & Erie opened in segments as they were completed: Akron to Cleveland opened on July 4, 1827. The first boat on the canal was the *State of Ohio;* the Wheeler brothers in Akron finished its construction on June 27. The Akron to Massillon section opened in 1828, Massillon to Dover in 1829, Dover to Newark in 1830, Newark to Chillicothe in 1831, and the entire canal from the Ohio River in Portsmouth to Cleveland on Lake Erie in 1832. From Cleveland, of course, passengers or cargo could travel two hundred miles by lake steamer to the Erie Canal in Buffalo, and from there east to the Hudson River and New York City. Eastern markets for Ohio produce now became an economic reality. That same year, the National Road from Baltimore (now Route 40), which had entered Ohio the same day the canal began in 1825, crossed over the canal in Hebron, Ohio, near Newark.

In 1833, 109 delegates decided to privately fund the Pennsylvania & Ohio Canal, often called the "Cross-Cut Canal," which would run from the Ohio & Erie Canal in Akron, Ohio, eighty-two miles eastward to New

Castle, Pennsylvania, where it would meet Pennsylvania's Beaver Canal, which connected it to the vast Pennsylvania Canal system, by then fully operational. Construction on the Cross-Cut started in spring 1835 and was completed by 1840. Afterward, Ohio canal boats could travel to New York via the Pennsylvania and New Jersey canals, an even shorter distance than via the Erie Canal.

The Sandy & Beaver Canal had been started in November 1834 with the intention of providing a more direct route to Pennsylvania from Ohio. It connected to the Ohio & Erie at Bolivar, Ohio, and extended seventy-three miles east to the Ohio River at Glasgow, Pennsylvania, just upriver from Liverpool, Ohio, and at the junction of Ohio, West Virginia, and Pennsylvania state lines. It was not operational until 1850, and it ceased general operation in 1852 when the summit level lost water supply.

In 1840, on the west side of Ohio, the twenty-mile Warren County Canal went eastward from Middletown, on the Miami Canal, to Lebanon. In the north, the State of Ohio extended the Wabash & Erie Canal from the Indiana state line to Toledo in 1843. By 1845, the Miami Canal had been built north from Middletown to connect to the Wabash & Erie at Junction, Ohio, thus becoming the Miami & Erie Canal, reaching from Cincinnati on the Ohio River to Toledo on Lake Erie, a full 301 miles.

In 1841, the Muskingum Improvement linked to the O & E Canal at Dresden and extended ninety-one miles south to Marietta on the Ohio River. In 1842 a twenty-five-mile feeder canal was completed, running northwest from the O & E at Roscoe to Brinkhaven, and in 1843, the Hocking Valley Canal was linked to the O & E at Carroll, and extended southeast fifty-three miles to Athens.

The State of Ohio's net cost for its canal network was about $13 million, but it was well worth it. Ohio was transformed from an undeveloped, wild territory to a productive, prosperous, enterprising state. Its interior was populated with immigrants in hundreds of towns, large and small, some of whom founded major national industries.

The thousands of canal boats were the public face of the canals after they were constructed and put into operation. The canal had been specifically built, of course, for them. The boats alluded to the romanticism of travel to far-off places, the transport of valuable and profitable cargo, and the adventuresome (sometimes brawling) nature of the crews. Many boats were named after towns along the canal where they were based, such as *Bolivar, Peninsula, Chillicothe,* and *Massillon.* Others were given names of famous Americans, such as *General Washington, Henry Clay,* and *Davy Crockett.* And some had names with more innovative origins, such as *Tidal Wave, Meteor, Albatross,* and *Flying Cloud.* Generally seventy to eighty feet long and fourteen feet wide, later boats were designed to carry heavy loads of fifty to eighty tons, but their flat bottoms drew only three feet of water. Two mules could easily pull them at the speed limit, about four miles per hour, a brisk walking pace.

The design of canal boats can be traced back to a 1752 boat built by Robert Durham, an engineer for the Durham Iron Works on the Delaware River in Bucks County, Pennsylvania, who needed a boat to ship his iron ore to Philadelphia. It was sixty feet long and eight feet wide, propelled upstream by poles and downstream with eighteen-foot oars. General George Washington and his troops used what were called "Durham boats" to cross the Delaware for a surprise attack on Hessian troops in Trenton on December 25, 1776. On December 1, 1776, Washington ordered Colonel Richard Humpton: "Particularly attend to the Durham boats which are very proper for the purpose."[2]

During the War of 1812, the fledgling U.S. Navy Department sent master shipwrights Adam and Noah Brown and a small group of ship's carpenters to build a fleet of Lake Erie gunboats on Presque Isle, near Erie, Pennsylvania, to defend against the many British vessels already there. Basic dimensions of these gunboats were within inches of early (circa 1825) Erie

Canal packet boats, which suggests that some of the Browns' two hundred workmen who remained in the Lake Erie area may have influenced Ohio & Erie canal boat designs. Like barns, canal boats were not normally built with plans but by experienced shipwrights who adapted wooden sailing-ship construction techniques to produce custom dimensions and designs.

There were infinite variations of boat design. Ohio freight boats had three small cabins with two open cargo holds between them. The captain and his family lived in the aft cabin, the crew in the bow and stern cabins, and mules or horses were housed in the middle cabin. Narrow catwalks over the cargo holds sometimes connected the cabins at the roofline for crew movement fore and aft. Other boats enclosed perishable cargo with roofs and sides and were sometimes converted for passenger transport as well, becoming combination passenger/cargo boats. Most boats were operated by lines, or companies, and were called "line boats."

Packet boats, built specifically for travelers, accommodated fifty to seventy-five passengers and were totally enclosed, with multiple curtained windows. Their interiors were divided into three sections, the largest being a lounge and dining area, which at night, with drop-down, triple-decker shelves as beds, became the gentlemen's sleeping compartment. The second largest was a private ladies' section with a washroom, and the third housed the crew and galley. These boats, called "through packets," traveled faster, at five or six miles per hour, and operated continuously day and night, with way stations every twelve to fifteen miles to get fresh teams. One such boat line advertised three days' travel from Cleveland to Portsmouth, 308 miles, including meals and lodging, for less than two cents per mile, but actual costs were much higher. Financial panics of 1837 and 1842 drove many lines out of business, and others revised operations to reduce costs.

Among the mix of designs in Ohio were "state boats" for canal maintenance; each of these had an open center deck and two cabins, one bow and one stern. The last traditional state boat in Ohio was built in 1909, at the W. J. Payne boatyard in Akron, the same place the first Ohio boat, the

State of Ohio, was built eighty-two years before. Some boats were floating commercial shops for sign-painters, printers, peddlers, et cetera, seeking business along the canal. In other states, boat designs evolved for specific cargo, such as the deep-hulled coal boats on the Lehigh Canal in Pennsylvania or the rectangular cross-sectioned hulls and flat-ended scows on the Chesapeake & Ohio Canal in Maryland. Each canal developed designs that became typical and unique to its region.

Canals became important parts of the countless small communities that grew up beside them throughout the state, providing needed goods and materials and opening up markets for local agricultural or minerals. Canal operations provided employment and transportation for many citizens and raised their standards of living through industry and commerce. The same was true in other states for all communities linked by canals. For the first time, the United States were truly united, able to compete industrially and commercially with European countries.

But by 1845, technology and industrialization, both inspired and enabled by canals, began to overtake canals. In the East, railroads were already competing with canals because they were cheaper to build and maintain, handled larger loads, moved much faster, and offered increasingly lower rates. They could operate all winter long, when canals were frozen. The Pennsylvania Railroad offered service from Philadelphia to Pittsburgh in 1852, and in 1857 it bought the Main Line Canal, which it used for freight only, from the state, after canals lost money for the first time in 1856. Railroad fever replaced canal fever. In 1861 coast-to-coast telegraph and railroads became some of the North's major weapons in the Civil War. By 1869, railroads had spanned the country, from the Atlantic to the Pacific, and there were forty thousand miles of railroad track in the United States, more than eight and a half times that of all canals. Railroads opened the far West, just as canals had opened the Midwest. As railroads expanded, canals gradually faded into obsolescence.

CHAPTER 3

Decline and Survival

Competition from railroads, loss of toll income, and the cost of maintenance ultimately made many canals unprofitable. Although faster steam-powered canal boats and "electric mules" (small electric locomotives) began replacing the slower four-legged mules, the wake caused by increased speed eroded canal banks and required more maintenance. Damage to banks from erosion and floods frequently stranded fleets of boats in dry canal beds for weeks. Across the entire system, canals slowly ceased operations because of flood damage or loss of water supply; in 1874, a major dam break rendered the Wabash & Erie Canal useless. As a result of reduced operations, fewer canal boats were built, and many existing boats deteriorated and disappeared into the mud of abandoned and empty canals.

New boats were built with much cruder and simpler design than the originals, and as these deteriorated they were replaced with cheaper and more functional rectangular scows used primarily for transporting bulk materials such as coal, stone, and lumber.

The decline and demise of canals was well documented by photography, which, in its infancy during the last quarter of the nineteenth century, captured life on the canal accurately. Early photographs show the classic canal boats deteriorating with time; unused boats were converted into floating houseboats or relocated on land beside the canal and treated as

These boats are stranded in a dry canal bed in Pennsylvania. William H. Shank, *Three Hundred Years with the Pennsylvania Traveler* (York, Pennsylvania: American Canal and Transportation Center, 1976), 80.

residences. Others show aging workboats repairing canals, always in need of maintenance.

As the canals saw less commercial use, they began to be used for recreation. In June 1886, this was dramatically illustrated by a group of prominent New York City citizens who cruised the Delaware and Lehigh Canals in a gravel scow converted into what they called a "yacht of the most approved canal pattern," named the *Molly Polly Chunker.*[1] Molly and Polly were the mules pulling the craft, and Mauch Chunk their destination, where canal boats that carried coal from the mines were called "Chunkers." Their scow was festooned with flowers, Chinese lanterns, lace curtains, and a striped yellow-and-black canvas awning. The group included, with family members, Robert W. DeForest (1848–1931), president of the Metropolitan Museum of Art; Henry Holt (1840–1926), president of Henry

Holt & Company, a publisher; Louis Comfort Tiffany (1848–1933), artist and inventor of Tiffany stained glass; Louise Knox (1851–1904), Louis's fiancée; and Walter C. Tuckerman, member of the American Geographical Society of New York. The object of the cruise was fun and experimentation with new dry-plate photography (developed in 1878) as they traveled by canal from Bristol to Mauch Chunk, Pennsylvania, the popular resort town where Tiffany and Knox were married the following week. The town boasted what is credited with being the first roller coaster, the Mauch Chunk Gravity Railroad, also known as the Switchback Railroad, built in 1846. It had not been used as a coal transporter since 1872; now the eighteen-mile, open-car railroad from Summit Hill to Mauch Chunk was a "must-do" experience and one of the most popular amusement attractions in America at the time. It operated until 1932.

From the 1880s into the early twentieth century, photographers captured many images of aging, classic canal boats overloaded with people in finery obviously enjoying Sunday church excursions or picnics. Other boats carried groups of well-dressed businessmen, apparently enjoying a company weekend getaway. It seems clear that canal boat rides had become a means of recreation, courtship, and celebration, much in the same manner as simple hay rides, horse-and-buggy jaunts in the country, and maritime cruises on yachts. Rural populations were relocating to urban areas to work in industry and commerce; for many, increased urbanization created a yearning to return to the natural, rural, and scenic countryside for relaxation and recreation.

The turn of the twentieth century was much more than a calendar change for America, which had become an international power that led the world in industrialization and mass production. It opened a new way of life: for instance, electricity and gasoline were now the sources of power, not horses, mules, or humans. By 1900, the country boasted two hundred thousand miles of railroads.

Captain Frank Lyons hosted this Methodist church picnic on his boat, the *Akron,* on the Ohio & Erie, near Cleveland, 1884. Postcard from the Pockrandt Collection of Historical Relics, Summit County Historical Society, published in Gieck, *A Photo Album of Ohio's Canal Era,* 244.

Automobiles and airplanes were new, rapid, long-range travel options, and in urban areas streetcars added further mobility. Automobiles, in particular, dramatically changed the culture of personal travel. With their advent, it became increasingly apparent that canal travel, along with horses and buggies, and hayrides were of a past generation, another century, another world.

Canals could compete only when they could carry enormous cargo loads. Accordingly, some existing canals were rebuilt or replaced by larger ones. For example, in 1900, the much larger Chicago Sanitary and Ship Canal replaced the Illinois & Michigan Canal, and from 1903 to 1918, the Erie Canal was completely reconstructed as the Erie Barge Canal, to accommodate barges carrying up to three thousand tons of cargo. The canal was 14 feet deep and 120 to 200 feet wide, with huge locks. Both of these canals remain in use today. In Ohio, starting in 1905, from Dresden to Portsmouth the Ohio & Erie Canal was abandoned, and the remaining portion from Cleveland to Marietta on the Ohio River was upgraded, until state funds ran out in 1909. Many locks were resurfaced with waterproof concrete.

For most of the original canals, in the face of railroad competition commercial use was impractical and too costly to even consider. Flood damage added to increasing maintenance costs, and in many cases, floods created such major damage to canals that they were abandoned because of failure of water supply. This happened to most Ohio canals in 1913.

By 1914 the Panama Canal was the only canal that aroused public excitement. Many original canals were inoperable. Others were converted into paved roads or railroads, returned to farmland or to nature, or built upon in urban areas. In 1918, the section of the Erie Canal that ran through the main streets of Syracuse, New York (the "American Venice"), was closed and covered over, creating Erie Boulevard. From 1920 to 1925, the Cincinnati downtown subway was constructed on the bed of the Miami & Erie Canal. Unused locks were dismantled for building materials or became overgrown. The last known "new" canal boat on the Ohio & Erie Canal was a replica of the original 1827 *State of Ohio,* built in 1925 in Akron to celebrate the centennial of the start of the canal. By the 1930s, when air travel became relatively common, train service became cheap and reliable, and improved roads encouraged long distance truck transport, the original canals were a lost chapter of American transportation history, along with the covered wagon and Native American trails that preceded them. The canal pioneers who built and used them were mostly gone. Canals had essentially become part of history, irrelevant to modern times.

Still, until 1924 some canals were operating commercially, among them the Chesapeake & Ohio and the Morris Canals. The Lehigh Coal &

Navigation Company operated the Lehigh and Delaware Canals profitably until 1931. That year, the Depression pushed the company into bankruptcy, and it sold forty miles of the Delaware Canal back to the State of Pennsylvania. In 1933, a private group called the Delaware Valley Protective Association was founded to protect the canal as a historic asset, and in 1940 it convinced the State to resume maintenance of the canal. The State of Pennsylvania acquired the remaining twenty miles of the canal, rewatered and repaired it, and named it Theodore Roosevelt State Park. Portions of both the Delaware and Lehigh Canals thus remained functional. Through the 1930s and 1940s, they were used for various recreational activities, including occasional nostalgic mule-drawn charter rides for community groups, such as in New Hope, where John Winter used an old scow for this purpose in 1942.

The Delaware & Raritan Canal operated until 1934, the middle of the Great Depression, after President Franklin Roosevelt had initiated possibly the most massive wave of national public works in history since the canal era to provide employment. The Works Progress Administration, under the National Recovery Act of 1933, created more than thirteen thousand federal projects, among them the electrification of rural America, dams, courthouses, tunnels, bridges, highways, streets, sewage systems, national parks, housing, hospitals, schools, universities, and work on the restoration of a few canals.

The most extensive canal was the Chesapeake & Ohio, abandoned since a 1924 flood. The U.S. government purchased it in 1938 and placed it under the care of the National Park Service (NPS), which planned to restore it as a recreational area. In 1939 the NPS hired young men between the ages of seventeen and twenty-five in the Civilian Conservation Corps to restore the first twenty-two miles of the canal, between Georgetown and Seneca. Each enrollee in what was organized as Company 333 received dress and work uniforms, was assigned a bunk in barracks, and received three meals a day and

$30 per month. They lived there year-round to provide for themselves and their families while they repaired and restored the canal, locks, and towpath, supervised by the NPS. In October 1939, workers in Hancock, Maryland, found the deteriorating hulk of ninety-three-foot long C & O freight boat No. 57, which the NPS measured and documented with scale drawings. The boat may have been the last commercial canal boat fully documented with measured plans. Although restoration work was discontinued because of World War II, in 1941 the NPS added a roof to a canal repair scow, called it the *Canal Clipper* after Bill Clipper, a canal boat captain, and began taking tourists on canal rides, towed first by a horse named Frank, and later, mules, through a restored section of the canal from Georgetown to Carderock, nine miles. This practice continued and became an institution.

World War II changed the priorities of Americans. The much anticipated "World of Tomorrow" of the 1939 World's Fair had become a reality. America became a world superpower with a booming postwar economy. The public, deprived of material goods for a decade and a half, wanted new cars, buildings, homes, roads, and technologies. By the mid-1950s, urban areas, some of which had begun as small canal towns, had expanded into large cities and suburbs, which—along with expressways, interchanges, and the interstate highway system—destroyed most old canal beds. All that was left of the once extensive canal system were a few state parks, mostly in rural areas, where an old lock and perhaps a section of an old canal bed with water survived. A few picnic benches and possibly a historical sign identifying the canal that once passed through were all that remained. For all practical purposes, the canal system was gone and forgotten. There remained only the latent seeds of canal restoration. In 1953, the Lehigh Coal & Navigation Company began to offer segments of its Lehigh Canal properties to municipalities for restoration efforts; Walnutport, Pennsylvania, was the first, and after 1964, it was joined by Weissport, Freemansburg, Easton, and segments between Allentown and Bethlehem.

The C & O Canal became an early postwar environmental lightning rod in March 1954, when associate Justice of the U.S. Supreme Court William O. Douglas (1898–1980), an avid environmentalist, opposed Congress's plan to convert the Chesapeake & Ohio canal bed and towpath in Georgetown into a motor parkway. Douglas challenged the editorial board of the *Washington Post,* which had supported the plan, to accompany him, thirty-six conservationists, and other newsmen, in a highly publicized eight-day, 182-mile walk on the canal towpath. The resultant national publicity caused the *Post*'s board to change its stance, and its support helped save the twenty-three-mile segment of the canal from Georgetown to Seneca. By 1971, the ill-conceived highway would be abandoned, and the restored canal segment designated a National Historical Park, saved for future generations. Every year since 1954, the Douglas hike has been reenacted under the auspices of the C & O Canal Association. Douglas participated in the event as long as health allowed, and in 1977, the park was dedicated to him.

The publicity resulting from the C & O Canal controversy (and its mule-drawn barge ride) may have inspired Pete Pascuzzo in New Hope, Pennsylvania: on September 18, 1954, he launched a pea-green scow, the *Mary P.* (his wife's name) onto the Delaware Canal. After his wife broke a champagne bottle at the launching ceremony before about a hundred people, two mules, John and Pete, led by driver John Winter, took a brief trip up the canal on the *Mary P.* with thirty adults, twenty-six children, and one boxer aboard. Pascuzzo soon renamed the craft and built three additional scows, making a fleet of four. Unlike the unpainted C & O barges, they were given flamboyant colors, and named the *American* (red, white, and blue), the *Liberty* (red), the *Independence* (blue), and the smaller *Spirit of New Hope* (orange). The scows carried fifty-five to eighty passengers and operated profitably for charter groups (four-and-a-half-mile trip) and day tourists (one-mile trip) from 1955 into the 1970s.

Disneyland may have been a possible source of inspiration for Pete (and

The *Spirit of New Hope* travels on the Delaware Canal in New Hope, Pennsylvania, 1967.

many others). The amusement park, which opened July 1955 in Anaheim, California, had been highly publicized by Walt Disney (1901–1966) on his weekly television show the previous year. Its attractions included Frontierland's "Mark Twain Riverboat," "Mule Pack," "Stage Coach," "Conestoga Wagons" and Fantasyland's "Canal Boats of the World." With Disneyland, Disney invented the concept of history and patriotism presented as entertainment, which had an enormous impact on its 50 million visitors over the next ten years. All mule-drawn canal boat rides since then have been in a similar spirit.

No such historically authentic canal boats had yet been built, but the seeds of nostalgia and national heritage had been sown. The next decade would see a realignment of national priorities and a growing awareness of air pollution and a deteriorating environment, which included miles of unused and neglected canal lands, as well as other historical landmarks being destroyed by developers. The federal government, and an aroused public, would begin to address these and other pressing social issues.

CHAPTER 4

Stark County, Ohio

The 1960s saw the assassination of national leaders, demonstrations against the Vietnam War, an antiestablishment youth culture, as well as the beginning of the space program. Civil rights, preservation of the environment, and women's liberation became popular political causes. In 1961, Jane Jacobs wrote *The Death and Life of Great American Cities,* which attacked postwar urban planning policies that were destroying traditional neighborhood communities. In 1962, Rachel Carson's *Silent Spring* raised public concerns about pesticides and pollution and fueled a growing national environmental movement. Betty Friedan's 1963 book, *The Feminine Mystique,* inspired the rise of women's liberation. The 1964 demolition of historic Penn Station in New York City aroused public outrage and inspired preservation policies. In 1967 the Summer of Love in San Francisco's Haight-Ashbury neighborhood brought national attention to a hippie youth culture that challenged traditional American values. The times were "a-changin'," and the abandoned canals would change with them.

In this tumultuous era of social change, the restoration and public reuse of canals would be achieved mainly through the quiet and diligent efforts of creative individuals and the collaboration of small community organizations. Their common motivation was to preserve and protect canals, the national heritage that had transformed America from a wilderness to a thriving nation of commerce, industry, and population growth. Over the following decades, such objectives received federal encouragement and support, and individual efforts expanded into a national movement.

The Clean Air Act of 1963 initiated study of air pollution problems. The Outdoor Recreation Act that same year charged the U.S. Department of the Interior with preparing a nationwide recreation plan and providing technical assistance to states, local governments, and private interests to promote the conservation and utilization of recreational resources. The Wilderness Act of 1964 implemented the National Wilderness Preservation System, which protected all federally managed land areas, including over 43 million acres under the National Park Service, which comprised 50 percent of all such federal lands. Human activities in parks were restricted to nonmotorized recreation, while logging, mining, building roads, and other forms of development were prohibited. The Motor Vehicle Air Pollution Control Act of 1965 set automotive emission standards. All this federal legislation provided many local environmental projects with essential public credibility and financial support.

Canal land development became part of this expanding movement embracing environmentalism and preservation. The State of Ohio had owned most of the canal right-of-ways since the canal's construction, and some scenically attractive segments of the canal bed had been turned into or became part of local state parks. Where feasible, some segments of the canal were rewatered and used for fishing or canoeing. Old towpaths in some areas were cleared to become scenic hiking or biking trails. In 1959, the State had rewatered a seven-mile segment of the Ohio & Erie Canal, from Canal Fulton to Massillon, in Stark County.

Then, in 1964, the State decided to divest itself of its canal lands. Under the law, government subdivisions such as towns and counties adjacent to canal lands would receive first priority; they could acquire the land free of charge, provided they would be used for public purposes. Private

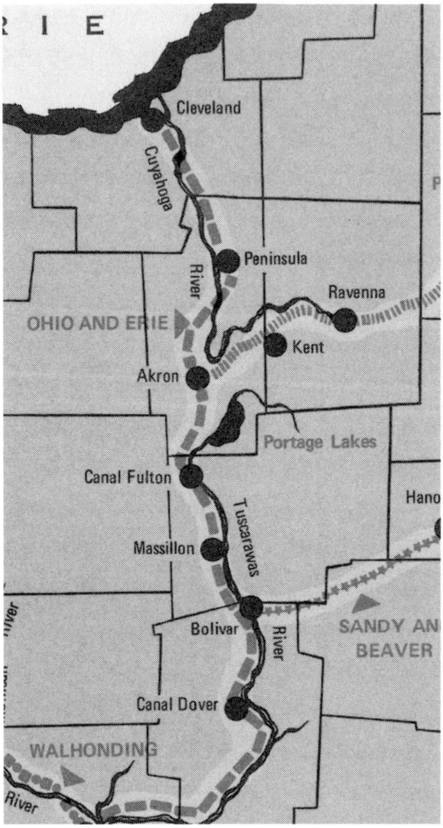

The northern section of the Ohio & Erie Canal, from Cleveland to Canal Dover, including Canal Fulton and Massillon in Stark County. Detail from the Ohio Historical Society's Canals of Ohio map. (See page 15.)

landowners, some of whom were already using part of the canal land without authority, would get second priority; they could now purchase canal land legally.

In Stark County, the three communities along sixteen miles of the old Ohio & Erie Canal—Navarre, Massillon, and Canal Fulton—were notified of this pending legislation, and mayors met to discuss options. There were some conflicting claims by Perry Township, between Navarre and Massillon, which almost killed the concept, but at a crucial moment, attorney Ralph S. Regula, Navarre village solicitor, then forty years old, suggested that the

Stark County commissioners take over all canal lands in the county outside of the three communities. It was his idea that the twenty-three miles of Stark County canal lands could be restored and renovated as a recreational hiking, biking, and nature trail. He argued that to allow private purchase of any canal land would prevent a continuous park, but if the land was owned by the county, continuity could be retained. This was the final proposal submitted to the county commissioners and to the state.

Regula realized that to create a twenty-three-mile Stark County park along the canal, community leaders and the general public had to be involved. Both financial and volunteer help would be needed. He remembered the 1954 success of Supreme Court justice William O. Douglas in saving the C & O Canal in Georgetown by taking a walk with the press. So he invited Canton conservationist Robert Vail Jr. and an experienced reporter for the *Canton Repository,* Al Simpson, forty-six, to accompany him on a sixteen-mile hike along the canal towpath from Canal Fulton to Navarre. Simpson's role would be to publicize the idea of a scenic canal park in Stark County.

The *Canton Repository,* Stark County's leading newspaper, published Simpson's major article, "Let's Save the Old Canal!" on May 10, 1964. Simpson described his hike with Regula along the O & E canal in a pouring rain. He compared the walk with that on the C & O Canal towpath by Justice Douglas and the press ten years earlier and wrote of the events leading up to the pending acquisition of the canal lands by Stark County. It was hoped that the County would establish a county park authority to develop the land, but Regula had much more in mind. He said on several occasions, "There's no limit to what could be done, absolutely no limit."[1] What Ralph Regula had in mind went well beyond Stark County. He envisioned a continuous 309-mile hiking trail along a restored towpath from Lake Erie to the Ohio River. Simpson urged local civic and welfare organizations to become involved in restoring the canal. He concluded:

"Stark County has an opportunity to set an example for other canal counties of the state, to lead the way in accepting responsibility for its section of the old canal and restoring it for the benefit of all Ohioans."[2]

Al Simpson wrote several more articles that summer, informing his readers of Stark County's potential by citing examples of communities that had restored canal towpaths and related facilities for recreational and tourist benefits. In June, Al, Ralph, and their families took a four-day tour of the C & O Canal in Maryland to see its restoration over the decade since Justice Douglas saved it from destruction. They met with Douglas and with officials of the National Park Service and the Department of the Interior, and toured the twenty-two miles of the restored canal. They saw the mule-drawn barge, the *Canal Clipper,* carrying tourists in Georgetown. Great Falls Tavern (built 1829-31), in Potomac, Maryland, on the canal bank near a lock, had been converted by NPS into a canal museum that attracted thousands of visitors annually. In Seneca, Maryland, a cycling trail on the towpath had been developed. The village sportsmen's club in Oldtown, Maryland, restored a five-mile stretch of the canal, including several locks, for hiking, biking, boating, and fishing.

Al and Ralph then explored every inch of the twenty-three miles of Stark County canal lands. By June 1964, the lands to be conveyed to Stark County had been described in detail, the concept of developing a recreational-historical park was articulated, and all was submitted to Columbus for state approval. Regula received encouraging remarks from Governor James A. Rhodes and director of the Ohio Department of Natural Resources Fred E. Morr. Naturalists and environmentalists backed the plan.

Al Simpson wrote a July 5 article describing seven potential community park areas along the canal, which could be linked together by a restored towpath, including Lock 4 Park near Canal Fulton. By August, the month in which the Beatles made their first triumphal tour of the United States, Pete Neidert, sixty-one, owner of the John Deere dealership in Canal Ful-

Pete Neidert, in his volunteer fireman's uniform, pilots his pontoon boat in Canal Fulton with family and friends at the village's sesquicentennial celebration, 1964. Seated passengers, from left: Kathy Neidert, Randy Neidert, Victoria Neidert, Bobby Sharon, Marilyn Gill, Emma St. Clair, Julie Sharon, Dorothy Schwendiman, and Jim St. Clair. Photo courtesy of Victoria Neidert-Hammer, Pete's granddaughter (seated front left).

ton, had built a motor-powered pontoon boat and skippered it two and a half miles up and down the canal to Lock 4 for visitors at Canal Fulton's sesquicentennial celebration.

In September, the State of Ohio officially deeded the Stark County canal lands to the county, and the Stark County commissioners appointed a Stark County Lands Development Advisory Committee, with Ralph Regula and Massillon's Mayor William Keen as cochairmen. Other members included John Cullen, Anthony Flex, Robert Gates, David Meyer, J. S. Sanders, Carl Weis, and Al Simpson, who served as secretary for the committee.

Ralph Regula spent summer 1964 talking to twenty-seven local civic organizations about canal renewal. These included Rotary, Kiwanis, and Lions Clubs, garden clubs, senior citizen organizations, landscapers' gatherings, Jaycees, and school and church groups. Previously a member of the Ohio Board of Education as a school principal, in November 1964 Regula was elected as an Ohio state representative, increasing his opportunity to move his canal restoration ideas forward.

On November 22, Simpson initiated a regular weekly column in the *Canton Repository* called "Along the Towpath," which he would continue diligently until March of 1970. It was intended "to chronicle the progress, report the obstacles, and in general, keep the residents of Stark County informed of developments 'Along the Towpath.'"[3] Simpson did so with reports of Boy Scout troop hikes, visits to obscure sections of the canal, skating events on the canal, the clearing of the towpath by a dozen relief workers supervised by canal foreman Paul S. Marks for the Stark County commissioners, the repairing of major leaks in the canal, and related stories and photos of the old canal from senior residents who recalled its last days of operation. He applauded the efforts of Ted H. Findley (1899–1969) of New Philadelphia, Ohio, president of the Canal Society of Ohio (founded in 1961), who advocated the creation of a park around Lock 2 in Akron to the city's planning commission. *Repository* readers were subjected to the weekly drumbeat of Simpson's columns, ensuring their continued interest and inspiring community action to restore the canal.

Simpson's January 24, 1965, article mentioned the mule-drawn scows that had carried New Hope, Pennsylvania, tourists on a watered section of the Delaware Canal since 1955. This provided additional inspiration for Simpson and led to his next step: He wanted a mule-drawn canal boat for Stark County.

On March 7, 1965, Al Simpson headed his column with the bold challenge: "Who'll Be First to Build Canal Boat?—Navarre, Fulton, or

Massillon?" His column had already reported rumors that other counties along the canal—such as Summit to the north and Tuscarawas to the south—were considering boat-building plans. He described the recent example of another mule-drawn sightseeing scow, the *Canal Clipper,* he had seen the previous year with Ralph Regula, operated by the NPS in Georgetown on the reconstructed Chesapeake & Ohio Canal since 1941. From May to October, the *Clipper* took visitors on a four-hour ride from Lock 3, through Lock 4, to Lock 5 in Brookmont, Maryland, although this operation would soon be suspended because of towpath damage.

For those unfamiliar with boat design, it may be helpful to explain some nautical terms. Anything that floats can be described as a "vessel." A "ship" is a large ocean-going vessel. A "boat" is a small ship. A "barge" is a flat-bottomed, shallow, rectangular box open at the top, generally used to transport bulk materials on water. It has no rudder, since it is usually towed or pushed by a boat. A "scow" is a barge with a rudder to control it when towed (i.e., by mules) or powered by engines. One observer described a scow as "a hole in the water lined with wood."[4] Both the Georgetown and New Hope vessels are scows, with added roofs to protect passengers from the elements.

A true "canal boat" follows traditional canal boat designs of the early nineteenth century, which generally had gracefully curved hulls with pointed bows and rounded sterns to move efficiently through water, and, like ships and boats, had enclosed structures, or cabins, that had sloping sides with windows, on above-water decks. Canal boats of the early nineteenth century were boats in the true nautical sense, which evolved from eighteenth-century ship designs.

Al Simpson wanted Stark County to be first with an actual authentic canal boat, not a modified barge or scow, and had suggested a competition between towns to encourage this result. Such a boat would be more accurately called a replica of a nineteenth-century boat. Responses were immediate. The Stark County Historical Center (now the William McKinley

Presidential Library and Museum), then headed by Gervis S. Brady (1913–1995), who promised help in "providing furnishings and equipment to give the boat an air of authenticity."[5] Craftsmen proffered time and tools to build a canal boat. Others offered wood for construction. In May, Pete Neidert offered a building at his dealership, Neidert's Tractor Service, on Erie Street, in which a boat could be built, just five hundred feet from the canal near Lock 4 State Park, a mile and a quarter down the canal from Canal Fulton.

Lock 4 Park was a perfect, scenic entry to the canal. In September 1938, during the Great Depression, the Ohio State Conservation Board allocated $25,000 to turn ten miles of the abandoned canal, from the feeder near Lake Lucerne north of Canal Fulton to Lake Street in northern Massillon, into recreational lands for hiking, camping, fishing, and hunting. This was accomplished with manpower (a $55,000 labor cost) from the Civilian Conservation Corps and expertise and equipment from the Stark County Engineers. The project included the construction of Lock 4 Park; the restoration of a functioning lock; and the building of a small simulated locktender's cottage beside Lock 4, using bricks made by the Metropolitan Brick Company of Canton, Ohio, that were scavenged from U.S. Route 21. "Baldie" Houck, of the Stark County Engineers, made the plans and supervised construction of the cottage.[6] By 1965, the gates now needed replacement, with the upper gate remains serving only as a dam. But the canal segment to town had been well watered since 1959 with a turnaround basin immediately above the lock. It was perfect for a canal boat ride, as had been demonstrated by Pete Neidert's pontoon rides in summer 1964.

On June 13, 1965, Al Simpson closed his weekly column with a cryptic paragraph: "And as of this week, there is renewed hope that Stark County can be the first canal county to launch an authentic packet since the early days of this century. I can't say more at this time. But a new development has encouraged me greatly. There soon may be exciting news 'Along the Towpath.'"[7] Something was up! Canal Fulton had just hosted the annual

Lock 4 State Park, Canal Fulton, 1967. The building is a replica of a locktender's house. The little girl in the picture is the author's daughter, Erika.

meeting of the Canal Society of Ohio, climaxed by rides on Pete Neidert's pontoon boat to Lock 4. In September, Pete surprisingly sold his pontoon boat to Clay Park for use as a lake launch. Perhaps he had a more important project on his plate?

The federal government continued to focus on the environment. In October 1965, Congress passed the Water Quality Act, governing water pollution in navigable waters, which included canals. That same month, at the urging of the First Lady, Lady Bird Johnson, it passed the Highway Beautification Act, regulating billboards and encouraging the planting of flowers in interstate medians.

Over the next year, Al Simpson's "Along the Towpath" columns focused on canal wildlife and canal history. He described the 1843 ride of former president and then U.S. congressman John Quincy Adams (1767–1848) on the eighty-three-foot-long Ohio & Erie canal packet *Rob Roy* for 232 miles from Cleveland to Columbus. He told of songs along the canal, bluebirds, and many other fascinating canal-related topics. On May 1, 1966, he reported that U. S. representative Charles A. Vanik (D-Cleveland), had proposed to the U.S. Department of the Interior the concept of making the section of the canal from Cleveland to Akron into a national park. However, no action from Washington was forthcoming. Al also mentioned that a "historic preservation recreation study" had been conducted on the old canal town of Roscoe, a section of Coshocton, where parts of the canal and triple locks were available for restoration.[8] Both of these developments were consistent with Ralph Regula's concept of a state-long parkway from Cleveland to Portsmouth, similar to the Chesapeake & Ohio Canal Park in Maryland.

Still, there had been nothing all year about the "exciting news" about a Stark County canal boat Al had promised in June 1965. Something was apparently up, but only Al and a few others knew what.

CHAPTER 5

Canal Fulton's Boat

The "exciting news" was finally announced in Al Simpson's column of June 26, 1966. Work on a canal boat had finally begun in Pete Neidert's farm equipment building in Canal Fulton. A few dedicated volunteers were laying out and cutting full-sized templates for a sixty-foot boat. They included Pete, who provided the power tools and truck for timbers; Paul R. Baird, seventy, president of Notnac Manufacturing Company, of Brewster, Ohio; Al Simpson; and Arthur P. Sweany of New Philadelphia, Ohio. For the last year, they had been working with the Stark County Canal Lands Development Advisory Committee to put together a plan of action inspired by volunteer donations and resources.

In June 1965, while attending the Canal Society of Ohio's annual meeting in Canal Fulton, T. H. "Ted" Findley, its founding member and president, had put the Advisory Committee in touch with Arthur Sweany. A flour millwright who had retired in 1955, he was eighty-six years old, had been a carpenter for sixty-two years, remembered the canal boat days, and had constructed several scale models of canal boats. He offered to draw a set of construction plans based on his memory of an actual boat, donate them to the Advisory Committee, and supervise construction of the boat. The boat in his plans was identified as an actual work boat, "*State Boat #5*, stationed at Canal Dover, Ohio, 1889," which suggested that this boat had been built when Mr. Sweany was about eight years old. When plans were completed near the end of the year, Sweany, Paul Baird, and others began to prepare the templates for the boat. Paul had offered to build it himself, but the committee believed that a volunteer civic organization would provide consistent and adequate manpower, so it continued to seek such a resource. More donations of green white oak trees were needed to make the hull timbers. But the fact that actual work had begun was a powerful attraction for the sorely needed additional skilled volunteer manpower and donated materials. It also attracted the public support of the Stark County commissioners, who said they would accept such a boat when constructed, and the pledge of Canal Fulton mayor John Cullen, who announced village council support. Al Simpson's column continued to promote committee needs, and people continued to respond generously.

There was more good news in 1966 for canal development. Commissioners of Tuscarawas County, Stark's neighbor to the south, followed Stark County's lead by applying for titles to all canal lands within its borders. In July, Canal Fulton held its first annual Old Canal Days Exposition, celebrating canal days, exactly 141 years after the Ohio & Erie Canal had begun on July 4, 1825, and in anticipation of the canal boat now under construction. Volunteers were diligently assembling canal boat materials. In July 1966, Bruce A. Crissinger Jr. donated two oak trees for the boat from his farm on Mudbank Road in Jackson Township.

In Washington, D.C., the National Historic Preservation Act of October 15, 1966, established several institutions: an Advisory Council on Historic Preservation, a list of National Historic Landmarks, a National Register of Historic Places, overseen by the National Park Service, and the State Historic Preservation Office, with fifty-nine officers, one for each state and additional ones for U.S. possessions. The act encouraged historical preservation on a national scale, which would grow in intensity over the next ten years as the bicentennial of the 1776 Declaration of Independence approached. The best news of all for the canal was the election in Novem-

Paul Baird and Roy Preece Sr. stand with oak logs to be milled for the canal boat project. Photo courtesy of John C. Harriman.

ber 1966 of Ralph Regula (R) to the Ohio state senate. The man whose vision inspired canal restoration in Stark County was now in a position to extend his dream to the entire state.

By December 1966 Paul Baird had located about thirty white oak logs and secured their donation to the boat project. Pete Neidert hauled them to the Nichter Lumber Mill in Canal Fulton for precision sawing and milling into timbers, after which they were dried in the lumber mill's kiln on Milan Street.

Simpson reported in his March 5, 1967, column about a jet-powered pontoon boat, the *Valley Belle,* which in 1965 was put into operation in Metamora, Franklin County, Indiana, on the Whitewater Canal. In just thirteen months, with a capacity of only thirty-five, it carried over forty-two hundred tourists through a lock. Al described Metamora as "a tintype community from the 1840s, alive and breathing in 1967. Preserved and restored, the canal runs through one of the most scenic sections of Indiana." Al saw Metamora as a model historical community for Canal Fulton to follow, and his vision was rapidly taking shape.

Indeed, by March, the stage was set with seasoned timbers, construction plans, and templates for a boat. All that was needed was a building crew. Al Simpson announced the solution to that remaining problem in his April 9, 1967, column in the *Repository.* A childhood neighbor of his, Jim Cozy (1916–2001), then president of the Louisville Sportsman's Club, twenty-five miles southeast of Canal Fulton, had called Al and offered his organization to build the boat. The group had two hundred members, including a number of skilled craftsmen, who were capable and ready to begin, and who would donate their time and talents to the Stark County boatbuilding plan. They could hardly wait! A canal-side construction site was selected just off Canal Street in Canal Fulton, opposite Parker Motors' garage, and just a short distance from the village square.

The small village of Canal Fulton was a perfect historical setting for canal boat construction and operation. It had been relatively untouched by urban development since its founding in 1826 on the Tuscarawas River, as the town of Fulton, by William Christmas and James Lathrop of Canton. Milan, an earlier town on the opposite bank of the river, had been founded in 1806; from the Treaty of Greenville in 1795 until the Treaty of Fort Industry in 1805, the Tuscarawas River had defined the western boundary of the United States. Delaware and Chippewa Indian tribes had still been numerous along its banks and used the river as a major north-south trail. By 1820, there were only twenty residents in Milan. In 1826, the *Canton Repository* advertised for five hundred day laborers to work on the canal for 30 cents a day and a jigger of whiskey, which, not surprisingly, brought more settlers into town. Fulton was on the O & E Canal section between Akron and Massillon, which was completed in 1828. Lock 4 south of Portage Summit was constructed a mile below Fulton and in canal days was known as "Fulton Lock" to boatmen, and "Mill Lock" to locals.

In 1853 Fulton merged with Milan, and combined they became Canal Fulton. The village had three hotels and seventeen saloons. In 1854, the Cleveland& Massillon Railroad reached the village, following the route of the canal. After 1865, it was well known as a mining town. Three hundred men worked in the hundred or so coal mines nearby or at the Fulton Mine Car Company. Many miners lived in or frequented notorious "Rogue's Hollow," which had five-to-seven-mile mule-drawn tramways connecting to the canal. With seven saloons, it was at its most notorious in the 1860s and 1870s. The intersection of Canal and Cherry Streets in Canal Fulton, with a saloon on each corner, was known as Brimstone Corner. By 1870 the Fulton Coal Company Railroad had connected mines directly to the canal basin, where coal was loaded onto canal boats, and in 1873, the town converted a warehouse into an Opera House that seated five hundred. By 1890, the population was 1,200. Coal shipments on the canal gradually diminished, while they increased on the railroad. The mines were depleted by 1905. After the canal ceased operations in 1913, the village survived as a farming community, and in the early 1940s it was a quiet, residential village of two thousand in a scenic, rural area, between industrial Akron and Canton, a bedroom community for both. The village had retained much of its Victorian architecture. It also had a secret weapon, Clyde Gainey, seventy-two, one of the last persons in Canal Fulton who remembered seeing working canal boats and who from 1919 to 1924 worked as an engineer for the Fulton Pit Car Works, founded in 1890. He was Canal Fulton village clerk from 1929 to 1956. After he retired in 1966 as a county engineer at the Akron-Canton Airport, he opened his personal Old Canal Days Museum in Canal Fulton, drawing seven thousand visitors each year. The twelve-by-sixteen-foot museum would become known as the Biggest Little Museum in the United States.

Jim Cozy and his club members got to work on the Canal Fulton boat immediately. Jim was a machinist in the engineering shops of the Hoover Company, in North Canton, ten miles southeast of Canal Fulton. Another Sportsman's Club member was Roy Preece Sr., forty-nine, who worked in Hoover's pattern shop and who now became construction chairman for the project. The club's first task was to provide Al Simpson with an illustration of its proposed boat to publicize in his weekly column. But all

Clyde Gainey's Old Canal Days Museum, Canal Fulton.

they had were the Sweany construction blueprints, which did not show the overall design or configuration of the boat. So on April 19, Jim and Roy called upon me. At the time, I headed Hoover's industrial design section, which designed the exterior appearance of all Hoover products and had often worked with Roy and Jim on Hoover design models. Familiar with both mechanical drawings and artistic renderings, I prepared an isometric illustration of the proposed Sweany boat based on plan measurements; this was the illustration Al Simpson published in his April 23 column.

Jim Cozy asked me to study the Sweany plans, since they provided inadequate critical information to builders, and asked me to prepare additional detail plans for construction. In studying them over the next week, I noticed that the plans were quite crude, many key details were missing, and some views did not agree with others. More significant, I noticed that the overall shape and proportions of the boat were inconsistent with photos I had seen of authentic nineteenth-century boats. To verify this, I visited Clyde Gainey's museum, where I found many actual 1875 to 1913 photos of Ohio boats. None resembled the Sweany plans. Compared to the square and angular cabins in Sweany's plans, the actual boats had more

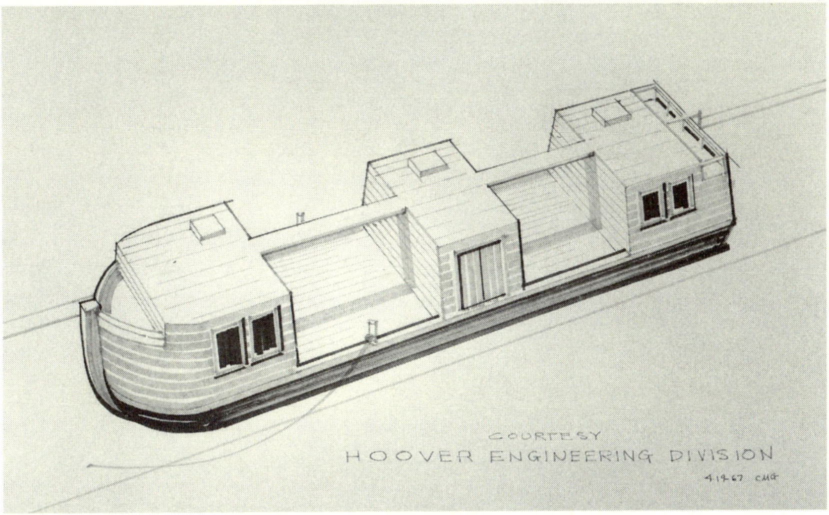

This isometric drawing of the Sweany boat design was taken from plan measurements.

subtle curvature in their cabins. Their bow stems rose almost vertically from the keel, unlike the long curved bow stem of the Sweany plans. I became convinced that the Sweany plans, while well-intentioned and based on the best memory possible, lacked the gracefulness, subtlety, and proportions of actual historical boats, and therefore did not meet the authenticity expected by all those who were involved in the project.

Faced with this dilemma, I decided to transform an authentic, historical photo of a canal boat into measurable dimensions and accurate proportions. I did this on May 3 through a reverse-perspective process. Technically, one can construct an accurate perspective view of an object from normal, three-view, dimensioned, mechanical drawings. I reversed this process, producing three-view, accurately proportioned and dimensioned mechanical drawings from a perspective photograph of a historical boat. The hull bottom, of course, was not visible in the photo. By May 16, I also had produced a large, full-color rendering of the canal boat that would

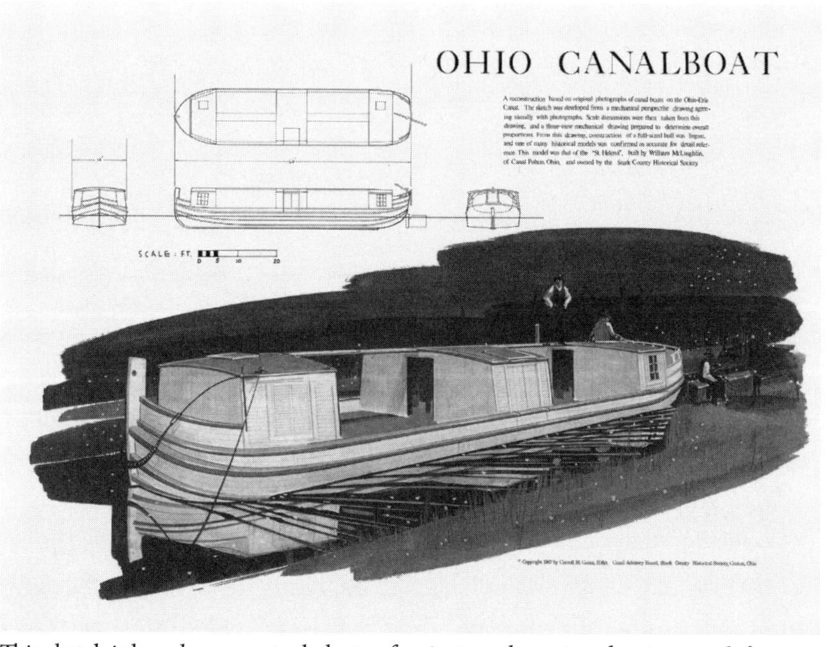

This sketch is based on an actual photo of a nineteenth-century boat; upper left are the dimensions/proportions derived from the sketch.

result from these drawings. This evidence convinced Jim Cozy and Roy Preece Sr. that the Sweany plans were historically inaccurate and I needed to develop new, more authentic plans before construction could proceed

By May 3, the construction crew had already erected six-foot-high old telephone pole segments on the work site to support a ten-by-ten-inch keel laminated from five two-by-ten-inch oak planks bolted together. The keel had been laid out and assembled at Neidert's Tractor Service. On May 25, they began to attach ribs to the keel to form a hull frame, which would be roughly sixty-five feet from stem to stern when finished. To make hull bottom planking easier, the frame was being built upside down. By June 2, I had developed critical new hull framing plans based on my studies, so that work progress would not be delayed. A volunteer crew got to work using the new plans, and included Jim Cozy; James Guest, thirty-four, IBM senior field engineer and vice president of the Louisville Sportsman's Club; Roy Preece Sr.; Paul Baird; Pete Neidert; and Clyde Gainey, as well as some new construction regulars—Walter Smith, sixty-five, owner of the four-floor water-driven mill in Crystal Springs built in 1925, and Gale Hartel, forty-three, general manager of the Massillon Automobile Club and owner of the Warehaus gift shop in Canal Fulton.

The crew worked Tuesday evenings and Saturdays. The Sportsman's Club created a canal boat display at the Central Plaza in downtown Canton, and sold "Canal Boat Boss" badges to raise funds. Offers of donated equipment and materials began pouring in from local businesses.

Louisville Sportsman's Club members lay the keel. Among them are Roy Preece Sr. (left, standing on bench with his hands on this side of keel), Jim Cozy (second from left, wearing derby), and Jim Guest (standing on a bench with his hands on the other side of keel). Photo courtesy of John C. Harriman.

I was busy that June searching for more authentic canal boat plans and more specific construction details. The hull bottom under construction, though later proven to be historically inaccurate, would be largely underwater and invisible, but the design of the boat's above-water portion would be highly critical for authenticity. The idea of tour boats operating in remaining scenic segments of canals was not a new idea; Al Simpson described many in his columns. I had seen several operating in New Hope, Pennsylvania, and Washington, D.C., but was dismayed that they were little more than modern barges or scows modified to accommodate tourists. I wanted to build, as did the Advisory Committee, an authentic historic replica of an actual nineteenth-century Ohio canal boat.

I sought original plans of such boats, particularly O & E canal boats. I contacted dozens of national museums and historical societies for authentic construction plans. I found a number of drawings showing the general lines of canal boats, for example, 1909 lines of a Lehigh Canal hinged coal boat, 1929 lines for a hinged Morris Canal freight boat, 1939 lines drawn from an actual C & O freight barge by the U.S. Department of the Interior, and even lines for Ohio freight boats. But I found no actual detailed, authentic construction plans. At Clyde Gainey's museum, I found a number of scale models of O & E freight boats, but their details were inconsistent, and they were small in size. The actual height of their cabins, if made full size, varied from five to eight feet. It was impossible to know which model dimensions were most authentic.

Then, I struck pay dirt at the Stark County Historical Museum. It was a huge, approximately five-foot-long, scale model of an Ohio & Erie canal freight boat that at first glance looked exactly like the full-color painting of such a boat I had made based on photographs. On the transom, sometimes called the "name board" of the model, was carefully painted: "St. Helena of Canal Fulton 1858–1933."

William McLaughlin's model of the *St. Helena,* in the William McKinley Presidential Library and Museum, Canton, Ohio.

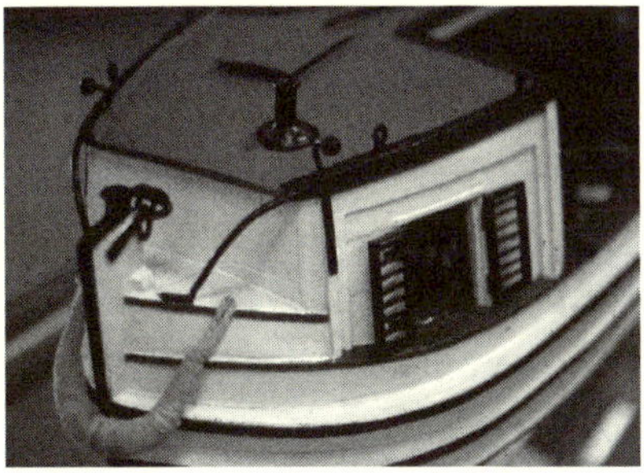

The model's bow cabin.

The model's aft cabin.

Later research revealed that William J. McLaughlin had built this model in 1933. He had worked as a ship's carpenter, repairing canal boats for his uncle Edward McLaughlin's business, the E. J. McLaughlin drydock, located on the canal a short walk south of Canal Fulton. Evidence of the drydock could still be seen there as earthen berms, just a few hundred yards from where volunteers were working. The drydock, established in 1830 by Michael Ruch, had been operated by the McLaughlin family, starting with James E. and followed by his son, Edward J. (born 1851), from about 1860 until the canal closed in 1913.

The McLaughlin model was incredibly detailed and in perfect scale, one inch to the foot, and at full size, would have been sixty-five feet, seven inches from stem to stern. Each plank and board of the hull, deck, and cabins had been fashioned individually. The model included interior

Above: William McLaughlin stands with his model of the *St. Helena.* Photo courtesy of McLaughlin Family Collection.
Left: McLaughlin's model of the *St. Helena,* tied up in the canal. Photo courtesy of McLaughlin Family Collection.

cabin details, such as bunks for crew, storage cabinets, and mule feed bins with sliding covers. Cabins had individual, removable roof hatches and shuttered windows. Iron fittings were everywhere, in perfect scale and function, including those for rope handling, deadeyes, docking rings, a towline cutter, rub rail facings, rudder mounts and chain, lantern fixtures, and a working capstan mechanism for raising and lowering the gangplank. There was even a canvas awning to cover the tiller deck to provide shade or protection against bad weather, as well as manual pumps for removing bilge water. Details included pike poles with hooks to fish the towline out of the water, a rope buffer, and even mule traces and singletrees. McLaughlin had obviously spared no effort to include every possible operational accessory of a working boat and was clearly familiar with every small physical detail. But was the model a scale replica of an actual boat?

Later, in a collection once belonging to the McLaughlin family, several fascinating photos were found. One, presumably taken in 1933, showed the seventy-five-year-old William McLaughlin standing proudly beside his model on the canal towpath. Another photo was of his model, floating in what appears to be the canal, and taken, probably also around 1933, from a low angle, to appear as an actual boat seen by a water-level photographer. The photo showed the stern of the model with its canvas awning extended over the tiller deck. On the transom, below the words "St. Helena of Canal Fulton 1858–1933" was text in small letters that had later been removed, as it was not on the Historical Society model: "J. Radcliff, Capt., Wm. McLaughlin, Bldr." This suggested a real boat with a real captain, perhaps one McLaughlin knew. The dates "1858–1933" probably referred to Mr. McLaughlin's birth year and the year he built the model.

Another photo was even more fascinating. It appeared to have been made from a dry-plate negative, which clearly dated it to the 1880s, and showed a stern view of an actual canal boat, virtually identical to the McLaughlin model, with two men and a woman on the tiller deck. On

This photo shows the nineteenth-century *St. Helena* of Newark, Ohio. Note the similarity of the name-board design to the McLaughlin model. Photo courtesy of the McLaughlin Family Collection.

the transom was written "St. Helena of Newark," referring to Newark, Ohio, the town where the O & E canal was begun in 1825. "St. Helena" possibly referred to the Catholic saint (circa 250–330 A.D.) or, more likely, the British island in the South Atlantic named after her, where Napoleon was imprisoned and died in 1821 and where slavery was abolished in 1832 (the same year the O & E Canal was completed).

Evidence suggests that William McLaughlin had based his 1933 model on an actual canal boat on which he had personally worked as a carpenter in his family drydock, and with which he was intimately familiar. What seems likely is that in his later years, when the original boat was either gone or deteriorating, McLaughlin had carefully created this model to preserve its memory in detail, in part as a memorial to his own career.

In measuring the scale model, I found that its major dimensions almost perfectly matched my reverse-perspective drawings. I was immediately convinced that this was an extremely rare historical find. The model was, for all practical purposes, a three-dimensional, detailed and accurate construction plan for an actual nineteenth-century Ohio & Erie Canal boat that had been repaired in Canal Fulton, the very place that twentieth-century volunteers were hoping to build a historical replica. If successfully built, it would become the first operational, historically authentic replica canal boat in the country since 1925. What an incredible coincidence, opportunity, and challenge!

Over the next three years, I measured every board and every fitting of the model and prepared mechanical drawings with detailed construction plans so that the volunteers could build an exact, full-size replica of it, including fittings, accessories, and color scheme (see pages 58–59).

There was only one problem. None of the volunteers or organizers had any idea how to construct a wooden boat. But we would learn, and somehow a genuine replica of the original *St. Helena* would be towed by mules on the canal where she was born!

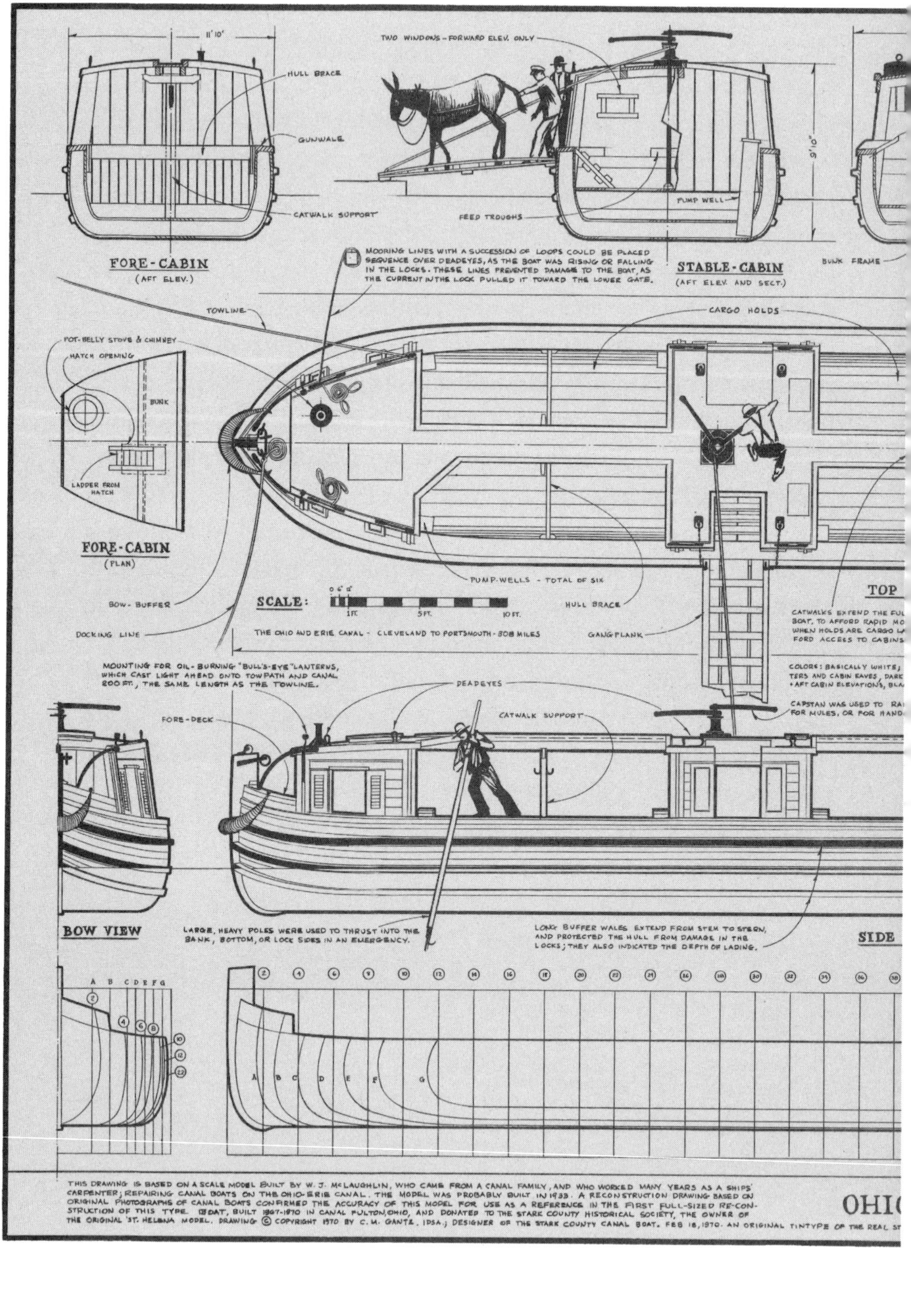
FORE-CABIN
(AFT ELEV.)
HULL BRACE
GUNWALE
CATWALK SUPPORT
TWO WINDOWS - FORWARD ELEV. ONLY
FEED TROUGHS
PUMP WELL
STABLE-CABIN
(AFT ELEV. AND SECT.)
BUNK FRAME
MOORING LINES WITH A SUCCESSION OF LOOPS COULD BE PLACED SEQUENCE OVER DEADEYES, AS THE BOAT WAS RISING OR FALLING IN THE LOCKS. THESE LINES PREVENTED DAMAGE TO THE BOAT, AS THE CURRENT IN THE LOCK PULLED IT TOWARD THE LOWER GATE.
TOWLINE
CARGO HOLDS
POT-BELLY STOVE & CHIMNEY
HATCH OPENING
BUNK
LADDER FROM HATCH
FORE-CABIN
(PLAN)
BOW-BUFFER
DOCKING LINE
PUMP-WELLS - TOTAL OF SIX
HULL BRACE
GANGPLANK
SCALE:
1 FT.
5 FT.
10 FT.
THE OHIO AND ERIE CANAL - CLEVELAND TO PORTSMOUTH - 308 MILES
TOP
MOUNTING FOR OIL-BURNING "BULL'S-EYE" LANTERNS, WHICH CAST LIGHT AHEAD ONTO TOWPATH AND CANAL 200 FT., THE SAME LENGTH AS THE TOWLINE.
DEADEYES
FORE-DECK
CATWALK SUPPORT
BOW VIEW
LARGE, HEAVY POLES WERE USED TO THRUST INTO THE BANK, BOTTOM, OR LOCK SIDES IN AN EMERGENCY.
LONG BUFFER WALES EXTEND FROM STEM TO STERN, AND PROTECTED THE HULL FROM DAMAGE IN THE LOCKS; THEY ALSO INDICATED THE DEPTH OF LADING.
SIDE
THIS DRAWING IS BASED ON A SCALE MODEL BUILT BY W. J. McLAUGHLIN, WHO CAME FROM A CANAL FAMILY, AND WHO WORKED MANY YEARS AS A SHIPS' CARPENTER; REPAIRING CANAL BOATS ON THE OHIO-ERIE CANAL. THE MODEL WAS PROBABLY BUILT IN 1933. A RECONSTRUCTION DRAWING BASED ON ORIGINAL PHOTOGRAPHS OF CANAL BOATS CONFIRMED THE ACCURACY OF THIS MODEL FOR USE AS A REFERENCE IN THE FIRST FULL-SIZED RE-CONSTRUCTION OF THIS TYPE BOAT, BUILT 1967-1970 IN CANAL FULTON, OHIO, AND DONATED TO THE STARK COUNTY HISTORICAL SOCIETY, THE OWNER OF THE ORIGINAL 'ST. HELENA' MODEL. DRAWING © COPYRIGHT 1970 BY C. M. GANTZ, IDSA.; DESIGNER OF THE STARK COUNTY CANAL BOAT. FEB 18, 1970. AN ORIGINAL TINTYPE OF THE REAL ST

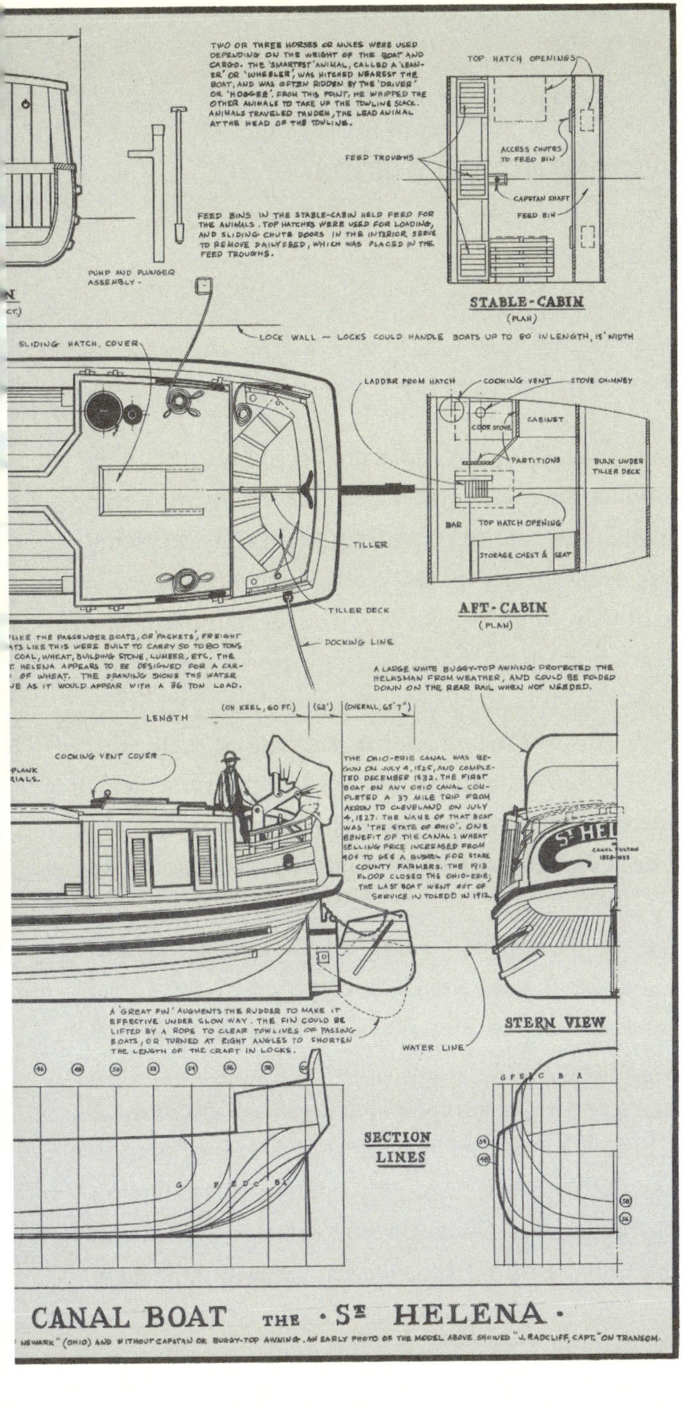

This plan was scaled from the McLaughlin model of the *St. Helena*.

CHAPTER 6

Construction Begins

In June 1967, all that existed on the construction site was the main keel, about fifty-seven feet long, resting on a half dozen six-foot-high telephone pole segments. Bow and stern patterns had been completed but were now replaced, using the new plans. Work, led by the Louisville Sportsman's Club and other volunteers, progressed on the framing during the summer. Builders cut bottom rib members from two-inch oak stock at the sawmill and bolted them to the keel, spaced two feet apart. They then bolted vertical side ribs to the bottom ribs and braced them with two-by-six-inch horizontal frame members that would support the deck. All were securely bolted together, resulting in a structural box frame to support the hull and deck planking. By July 9, fourteen rib sections had been completed.

That month, the second annual Old Canal Days Exposition in Canal Fulton featured Pete Neidert's pontoon boat, operated by the Canal Fulton Rotary Club under the supervision of Gale Hartel. The craft took twenty-five passengers on forty-minute rides, one and a quarter miles down the canal to Lock 4 and back, during which they were entertained by a narrative of canal history conducted by a club member. The pontoon operated weekdays on a charter basis, and was open to the public on weekends throughout the summer. Under the editorship of Gale Hartel and Clyde Gainey, the Canal Fulton Exchange Bank began publishing

Sportsman's Club members added ribs to the keel. Roy Preece Sr. (left) and Clyde Gainey stand on the ground. Photo courtesy of John C. Harriman.

This book's author (right) reviews plans with Roy Preece Sr. (left) and Jim Cozy (center) at newly completed stern post. Photo courtesy of John C. Harriman.

the *Canawler,* a tabloid that featured Ohio canal history and progress on the boat construction, " as a public service . . . so that this generation and future generations . . . will remember the local historical places and events that have helped to give us this great nation."[1]

In his July 23 column, Al Simpson reported about currently operating canal boats, though none was as authentic in design as Canal Fulton's. In 1966, at the Great Falls Tavern in Potomac, Maryland the National Park Service put into operation a new forty-by-eleven-foot mule-drawn scow

A National Park Service scow, the *John Quincy Adams* (*Canal Clipper II*) at Great Falls Tavern, Potomac, Maryland. Photo taken 1972.

named the *John Quincy Adams* (after the man who turned the first ceremonial spade of dirt on the C & O Canal in 1828), also known as *Canal Clipper II,* because it replaced the twenty-five-year-old *Canal Clipper.* The new scow carried visitors for a half mile, through Lock 20. Unfortunately, it would be lost in the 1972 Tropical Storm Agnes. Al also mentioned the Delaware Canal in New Hope, Pennsylvania, where tourist scows had been operating with mules since 1955, and the *Valley Belle* pontoon boat ride in Metamora, Indiana, on the Whitewater Canal. Al reported rumors that other replica canal boats were being planned in Ohio and New York but said that Stark County's boat was further along than any of those. By November, the Canal Fulton boat framing was essentially finished, resembling a giant prehistoric dinosaur skeleton. It was carefully wrapped in sheet plastic to protect it from the elements over the winter.

Meanwhile, the community of Canal Fulton was organizing its resources to maximize its contribution to the boat project. That winter, Ed

Stern end of the boat frame, November 1967.

The boat's frame was protected with plastic sheet during winter 1967–68.

Harriman, forty-six, president of the Exchange Bank of Canal Fulton, and Gale Hartel, attended a governor's conference on tourism to get ideas on promoting the canal revitalization. Both men became de facto community leaders, coordinating support for the town's ambitious project. They published the second issue of the *Canawler,* and Harriman was elected president of the Rotary Club, a steady resource for construction manpower. The Canal Fulton Heritage Society was incorporated to provide organizational, promotional, and volunteer support for the project.

The plastic was removed as work on the hull began again on April 30,

1968, the month of Martin Luther King's assassination and President Lyndon Johnson's signing of the Civil Rights Act of 1964. In preparation for planking, the frame was coated with red lead preservative primer paint to protect it from deterioration in water. In May, using two pontoon boats, the Rotary Club resumed sightseeing operations on the canal on Sundays and

Paul Marks's plank steam box stands in the foreground.

The hull planks are attached to the boat's frame.

holidays. On June 5, the news of Robert Kennedy's assassination shocked the nation. The Stark County Historical Society had hired Paul Marks to work full time on boat construction, and by mid-June he had built a steam box from a fifteen-foot section of a sixteen-inch diameter steel pipe, donated by the East Ohio Gas Company. He welded one end shut with a steel plate, propped the open end up on a sawbuck about three feet high, and at the lower end built a masonry firebox, fired by two oil burners. He covered the firebox with sand, to retain the heat, then filled the large pipe with water and began boiling 5¼" x 1¼" oak planks for the hull, sawn and especially milled for easy caulking. After several hours in the boiler, the planks became highly flexible, and could be bent while hot to fit the many curved sections of the ribs on the hull and attached to them with chrome-plated, spiral-twisted, five-inch steel spikes. Where one plank end met another, an oak plate was placed inside the hull and bolted to both plank ends.

The planking went smoothly on the sides of the hull, where planks were relatively straight, but where severely bent planks met the stem post (the prominent, vertical, wooden bow piece rising from the keel), the volunteers faced a serious challenge achieving a watertight seam. Help came from a local North Canton sailboat owner, Bill Embly. He recently had a sailboat built by a sixty-one-year-old master shipwright, who in Lloyds of Maryland built vintage wooden sailboats known as Chesapeake Bay Skipjacks. I asked Bill to tell the shipwright about the Canal Fulton boat and ask him for his help in our construction dilemma.

James B. Richardson (1907–1991) was already a living legend. "Mr. Jim" was descended from a line of master shipwrights on the Eastern Shore that traced back to pre-Revolutionary days, and he still spoke in a seaman's dialect that was distinctly eighteenth-century British. Not only had he built many Skipjacks, but at that very moment was building the *Adventure,* a historical replica of a seventeenth-century ketch to commemorate the Tercentennial founding of Charleston, South Carolina, in 1670. The boat was to be launched in 1970 at Charles Towne Landing, the site of the original landing by the first settlers, and would remain there until it was replaced in 2008 due to deterioration. (Richardson had also built historical ship construction replicas for the Smithsonian Institution in Washington, D.C.) James Michener would immortalize Richardson in his 1978 book, *Chesapeake,* as the human model for the individualistic, devout, Paxmore family boat builder. A more experienced, skillful, and colorful shipwright could not have been found.

Jim agreed to take a look at the canal boat hull in Canal Fulton and arrived there on June 26 with his son-in-law and his adze, a special axe used to hew timbers. He used it to carve out a channel in the stem post to accept the boiled planks, and, with the volunteers, he applied a few planks to demonstrate the process. He also installed a two-by-four-inch sub-keel that curved around and completed the difficult stem assembly. At an evening meeting in North Canton at the Harleigh Inn, he advised us volunteers on the caulking processes, materials, and tools we would need for the next step. Al Simpson asked him what he thought of our work so far. He responded, "Well, I'm very pleased with what you have done. I'm amazed, however, that you all would take on this much of a boat as your first venture. You know, this is rather an ambitious undertaking. But I think it is terrific! Properly cared for, though, I think she ought to last 50 years."[2] (Years later, many would regretfully remember that "properly cared for" part!) Paul Marks and volunteers continued the hull planking process, and by August, the hull was two-thirds planked, and covered with wet burlap to keep it from drying out.

While volunteers remained at work on the canal boat, national events continued to shock the country. On August 28, street riots broke out in Chicago at the Democratic national convention in protest of the Vietnam War. The country seemed to be coming apart at the seams.

In September, the Louisville Sportsman's Club relinquished its leadership

Left: Paul Marks (foreground) watches Jim Richardson attaching the sub-keel.

Below: The crew bends and attaches a plank to the hull. Among them are Jim Cozy at top right, Paul Marks bottom left, with cap, and Walter Smith behind bow. Photo courtesy of John C. Harriman

role, since there were more than enough local volunteers. As the designer, I became the de facto construction supervisor and began calling for volunteers to build a plastic greenhouse over the hull, so that work could proceed during the winter. By November, when Richard Nixon was elected president, we had built a sixty-five-foot by sixteen-foot wooden house frame covered with translucent sheet plastic, and with gas-powered portable heaters, or

Here, the winter greenhouse has been built over the hull and the steam box fired up for plank bending.

Tapered planking on the stern of the hull.

salamanders, donated to heat the interior, it became a most comfortable and well-lit working environment.

As reported by the Ohio Historical Society that November, the excitement in Canal Fulton was part of a larger canal craze developing throughout the state.[3] A million-dollar plan was being developed in Coshocton to restore a mile-long canal section and float one or two canal packet boats and to restore the canal town of Roscoe Village. In western Ohio's Piqua Historical Area, there was a $75,000 project to restore a mile-long section of the Miami & Erie Canal, on which a canal boat would be built and operated. Lock restorations were planned in Butler County and along the old Wabash & Erie Canal. Canal Fulton now appeared to be in an Ohio horse race to build the first replica canal boat.

Planking of the stern was the most difficult task at hand. Each plank had to be not only boiled and bent but custom-tapered to create a hemispherical hull shape leading to the transom, the highly visible name-board of the boat. North Canton professional carpenter Richard Mohler, forty-two, and his apprentice, David Wahl, were hired for this complex project, and they began work in December 1968. Just as the planking was nearing completion on Christmas Day, *Apollo 8* circled the moon, sixty miles above its surface, with astronauts Frank Borman, James A. Lovell, and William A. Anders aboard. Americans rejoiced in this major step toward President Kennedy's stated goal of landing on the moon within the decade, while we canal boat volunteers were inspired, knowing that our own difficult goal was also progressing at good speed.

Next came the painting of the entire hull with red, algae-resistant bottom paint. The major volunteer task for the winter was manually caulking over a half mile of hull seams. Retired expert caulkers from Lake Erie's Port Clinton, Ohio, Tony Patz and Laurence Johannsen were hired to spend five days in Canal Fulton in January 1969, to demonstrate caulking seams and educate volunteers in the unique techniques and materials required.

Prior to their arrival, special caulking tools had to be made by hand and caulking materials purchased. The first step of the process used oakum,

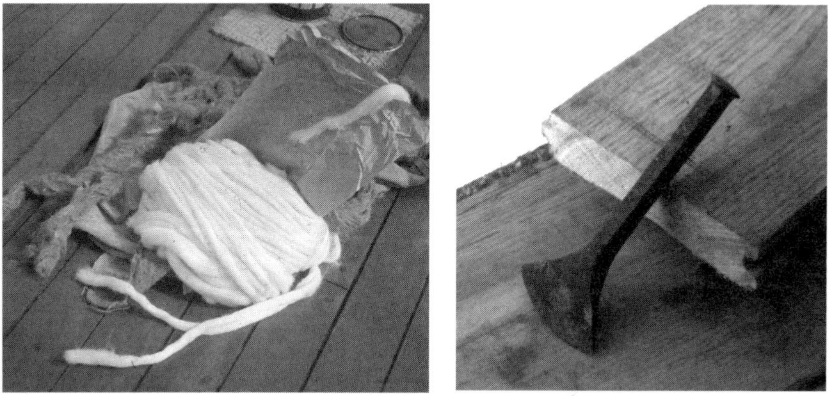

Left: The hull was caulked with oakum and cotton. Right: A caulking iron.

Volunteers apply seam compound to the hull, over the caulking.

The rub rails, attached to the hull.

a loose, slightly tacky, ropelike, hemp fiber, which was driven into the seams with a mallet and caulking iron so that it filled the seam to about one inch below the plank surface. On top of the oakum, a thick strand of white cotton caulking material was driven into the narrow seam, again with a mallet and caulking tool, until it was about a half-inch below the plank surface. The process was a laborious one, proceeding inch by inch and requiring a delicate touch. These materials were well saturated with linseed oil, and the remaining half inch of the seam was filled with white, puttylike Kuhl's seam compound #2, which would remain flexible when the oak planks swelled with water. Finally, the entire hull was painted with a second coat of copper-based bottom paint.

The heated greenhouse provided comfortable working space for the large number of volunteers who participated in the caulking process, and by March 1, the hull was pretty well caulked. Dick Mohler and his assistant had completed the complex stern planking and now began building the three longitudinal canal boat rubbing strakes on the hull exterior, to protect the hull from scrapes in the concrete lock. They would be later faced with ⅛" x 3" iron bands for durability.

The Stark County Historical Society appointed a Canal Branch Board of Managers, chaired by Ed Harriman, which included vice chairs Gale Hartel and James Guest and volunteers Paul Baird, Roy Preece Sr., Pete Neidert, James Cozy, Clyde Gainey, Walter Smith, and myself, as well as other contributors: Al Simpson, Ralph Regula, Ronald K. Bennington, Mrs. Mark Herbst, Mary Muhlhauser, Robert Christensen, Thomas E. Myers, Robert Seiple, Gary Stephan, and Mrs. George Wincek. Ex officio members were Gervis Brady, executive director of the historical society, and Dr. R. K. Ramsayer, president. The first meeting was April 29, at which members began planning for all that would be required for boat operations, including hiring a crew, building a ticket office, acquiring mules, and advertising and publicizing boat rides. On May 2, 1969, the construction volunteers formally donated the boat to the historical society, which had agreed to put the boat into public service when completed.

By May 31, the hull was essentially complete with rubbing strakes, and by June final painting was completed. The ultimate challenge became how to lift the sixty-five-foot-long, twelve-ton hull, carefully invert it, and place it gently into the canal. This highly anticipated event was being called the "splash-in," a term borrowed from NASA's Apollo spacecrafts, which landed in the ocean. (In fact, on May 23, *Apollo 10* had achieved lunar orbit, and on May 26, it splashed down in the Pacific. Incredible photos of the earth from space reminded the world that the planet was a unique island of natural beauty that had to be protected and preserved.)

July 12 was scheduled for canal boat splash-in during the third annual Canal Fulton Old Canal Days Exposition. In June, in preparation for this event, Paul Marks and his crew installed in Lock 4 four brand-new, operational, white oak lock gates, which they had built from scratch, using patterns made from the old, rotted gates. Paul Marks, Paul Baird, and Pete Neidert had found the oak trees and had them sawn to size at Nichter's sawmill. The huge cylindrical wooden hinges were a problem: no lathe

The new lock gates—the temporary white wooden bridge over lock would be removed for operation.

big enough to turn them could be found. But Robert W. Raper, president of Akron Extruders, provided one at his Canal Fulton plant, as well as the labor to do the job. The two upper, smaller gates weighed three tons each, and the lower, larger ones, five tons. Paul Marks and his crew also repaired and painted the concrete-surfaced lock itself. At the time, it was hoped that the boat could be taken through this, the only operational lock of the original 146 on the main canal. On June 20, Dick Mohler completed the stern transom on the boat.

Canal fever had intensely gripped the community. After all, it was to be the first launching of an actual new canal boat in this canal in a generation! Local television and press promoted the anticipated splash-in. Two pontoon boats were ready for action to take visitors down the canal. On July 5, the plastic winter boathouse came down. The hull was complete

A July flood nearly filled the construction site. Photo courtesy of John C. Harriman

and ready to be rolled over and launched. But in a few days, the boat was unexpectedly deep in the water, way ahead of schedule. Several intense rainstorms had dumped record amounts of water into the river and the canal, flooding the construction site as well as many parts of the village. Those who had witnessed both said it was the highest water level seen in the village since 1913, when a similar flood destroyed the canal. Canal water reached the bottom planks of the inverted hull. It took a week for the water to drain from the site and another week for it to dry out enough to get heavy equipment onto it.

So the Old Canal Days festival took place on schedule, July 11, 12, and 13 without the planned canal boat splash-in. On July 16, aboard *Apollo 11,* Neil Armstong, Buzz Aldren, and Michael Collins were launched. Armstrong and Aldren landed on the moon July 20 in a lunar module while Michael

Collins waited in *Apollo,* which returned to earth and splashed down in the Pacific on July 24.

Meanwhile, the canal boat splash-in had been delayed until July 26, when the construction site had dried out. Still, the coincidence of the almost simultaneous completion of both these long-anticipated achievements was uncanny. The splash-in date drew a crowd of more than two hundred, including Ohio congressman Ralph Regula and *Repository* reporter Al Simpson. A local contractor, Henry A. Selinsky of Canton, donated the use of two sixty-ton building cranes and crews to operate them. They inched into the narrow, still muddy, construction site and were positioned over the hull, each with a large, reinforced, rubber-enclosed, cable sling positioned around and below the hull, so that the twelve-ton dead weight was distributed evenly fore and aft on the two slings.

After the flood waters receded, Henry A. Selinsky's cranes lift and rotate the hull. The man in the foreground with the hat is Paul Baird.

The hull rolls over, exposing temporary bracing.

Happy volunteers, from left: Carroll Gantz, Dick Mohler, Roy Preece Sr., and Pete Neidert.

Slowly, the cranes lifted the enormous hull a few feet above the ground, and the slings were very slowly rotated in place, rolling the hull over until it was upright. The crowd was deathly quiet with expectation and apprehension while the hull was slowly turned over, like a pig on a spit. It creaked and flexed almost like rubber with the movement, even though it had been reinforced with temporary diagonal bracing timbers to keep it structurally rigid. It was swung slowly and ponderously out over the canal by the cranes and gently lowered into the canal. The slings were carefully disengaged and, at last, the hull was afloat at exactly 2:35 P.M. The crowd on the canal banks cheered loudly as "she" (now an actual floating boat) was drawn to the shore and moored to a tree on the construction site. A major milestone had been achieved, and the hundreds of volunteer workers who made it possible sighed with a combined sense of accomplishment, relief, and anticipation of finishing their challenging task.

The summer pontoon boat rides demonstrated that there was strong public demand for a scenic canal boat ride. By the end of the 1969 season, three thousand residents and visitors from twelve states and thirteen foreign countries had taken advantage of the Canal Fulton Rotary Club's pontoon boat tours down the canal, indicating that a larger authentic boat with a working crew could attract even more potential visitors to Canal Fulton.

The date for launching the boat and beginning of operations had already been set for July 1970, only a year ahead, so there was no time to lose! The boat was only half done, but all went back to work with renewed enthusiasm and dedication.

CHAPTER 7

The *St. Helena II* Is Born

For several weeks, water leaked in and filled the bottom of the hull, causing some of us apprehension, but soon the planks swelled, and after an electric bilge pump emptied it, the hull became as tight as a drum, thanks to the expert caulking job. Temporary bracing was then removed from the hull. In August 1969, as soon as they dried out, interior hull timbers were coated with red lead paint. This month was filled with news reports of the gruesome Charles Manson Family murders in California (August 9), and the Woodstock festival in upstate New York (August 15).

Over the next month, a wooden framework was built directly onto the hull and covered with plastic sheeting. By October, a floating greenhouse with an efficient space heater was ready, allowing construction throughout the winter. The hull was now ready for its final transformation into a replica canal boat.

The final phase of construction called for the planking of the deck; building three cabins and furnishing their interiors; roofing the cabins; building gunwales, catwalks, transom, tiller and tiller deck, and rudder assembly; forging assorted custom iron hardware; and painting everything. North Canton contractor Dick Mohler began preparation work in October 1969 by having wood cut and milled from the oak logs at Nichter's sawmill. Deck planking and heavy cabin framing timbers were sawn and milled to shape. Deck planks were of 1½" x 6½" tongue-and-grooved oak. Most of

Jim Guest and his daughter Cindy survey the bracing removed from the hull and the bilge pump works to empty water that had leaked into the boat.

The winter greenhouse is again ready for work.

the vertical oak cabin timbers were rough sawn to four by six inches but for the horizontal rafters, four-by-twelve-inch timbers were required, so they could be shaped to match the curvature of the roof. Dick Mohler and others worked to my exact plans, which included five hundred individual components and accessories, including hatches, hardware, ladders, pike poles, shutters, et cetera, all scaled from the McLaughlin model of the *St. Helena* at the Stark County Museum.

The main deck and gunwales are under construction.

By January 1970, Dick Mohler began actual construction work on the boat, starting with the main deck and gunwales, and by the end of March these were both essentially done. The three cabins had all been framed with heavy timbers. They were to be left in rough-sawn, unfinished wood. Crew had traditionally occupied the forward cabin, and the captain and his family the aft cabin, with its rear access stairway to the tiller deck. Hatches on the fore and aft cabin roofs allowed the crew to climb on top, where they could walk the length of the boat on catwalks, over the two main cargo holds between cabins. Decks were now constructed in these holds to provide seating space for sixty passengers and to allow them to walk the length of the deck to all cabins. The middle cabin traditionally accommodated the mule "power system," so this cabin featured feed bins and compartments for grain but would also become the primary boarding point for passengers.

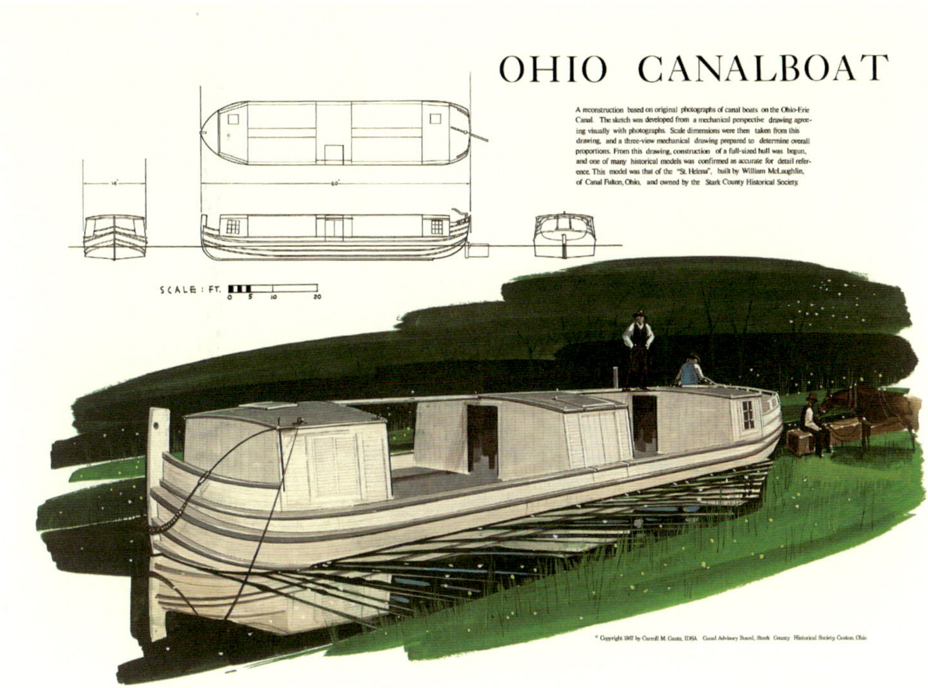

This sketch is based on an actual photo of a nineteenth-century boat; upper left are the dimensions/proportions derived from the sketch.

Navigable rivers (blue) and canals (magenta) of the United States, 1905 (Albany, New York: Brandow Printing Company, state printers, 1905), color added by author.

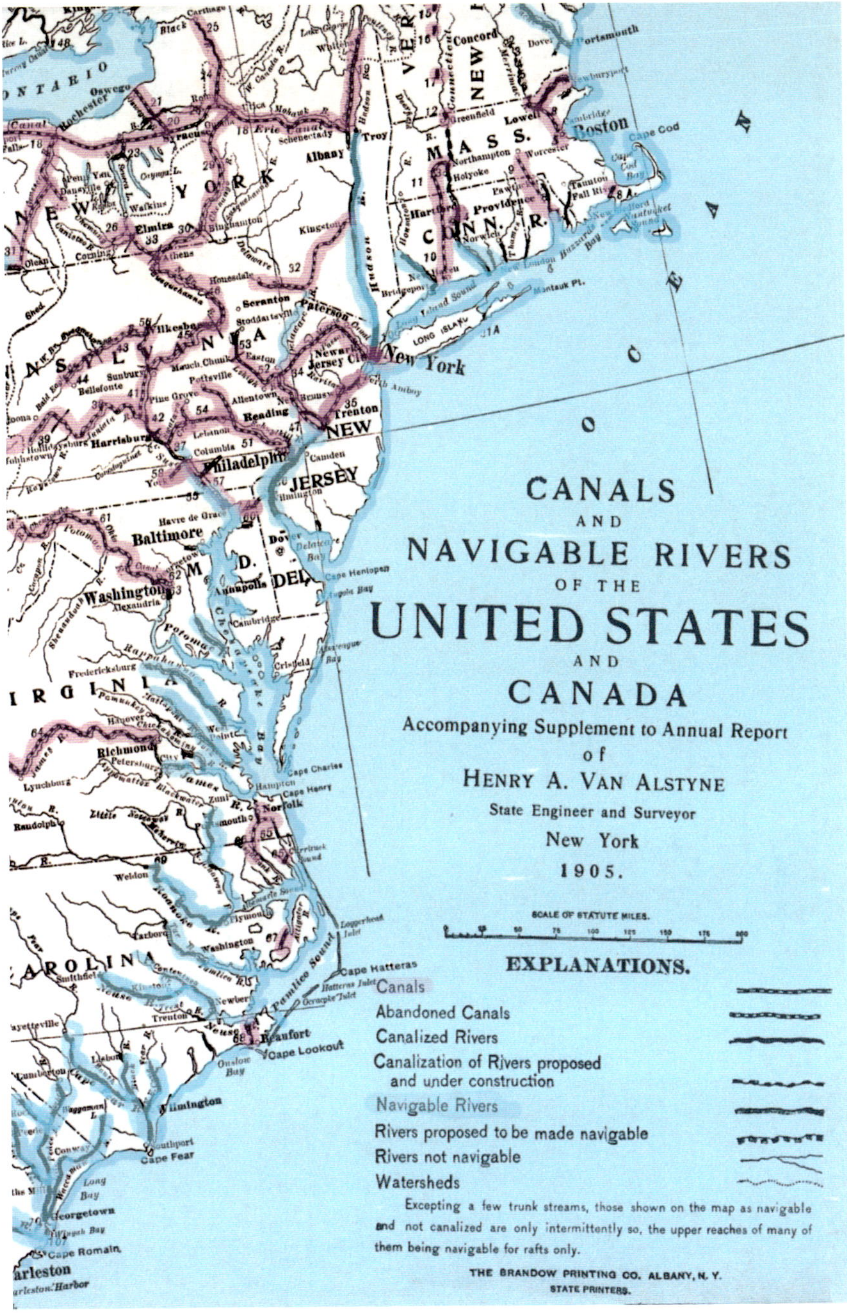
CANALS
AND
NAVIGABLE RIVERS
OF THE
UNITED STATES
AND
CANADA
Accompanying Supplement to Annual Report
of
HENRY A. VAN ALSTYNE
State Engineer and Surveyor
New York
1905.
SCALE OF STATUTE MILES.
EXPLANATIONS.
Canals
Abandoned Canals
Canalized Rivers
Canalization of Rivers proposed and under construction
Navigable Rivers
Rivers proposed to be made navigable
Rivers not navigable
Watersheds
Excepting a few trunk streams, those shown on the map as navigable and not canalized are only intermittently so, the upper reaches of many of them being navigable for rafts only.
THE BRANDOW PRINTING CO. ALBANY, N. Y.
STATE PRINTERS.
ONTARIO
NEW YORK
MASS.
CONN.
R.
NEW JERSEY
VIRGINIA
CAROLINA
Boston
Albany
New York
Philadelphia
Baltimore
Washington
Richmond
Norfolk
Cape Cod
Cape Hatteras
Cape Lookout
Cape Fear
Charleston
O C E A N

The crew bends and attaches a plank to the hull. Among them are Jim Cozy at top right, Paul Marks bottom left, with cap, and Walter Smith behind bow.

Volunteers apply seam compound to the hull, over the caulking.

Henry A. Selinsky's cranes lift and rotate the hull. The man in the foreground with the hat is Paul Baird.

Dick Mohler does janitor work in the partially framed middle cabin. The catwalk above him connects cabins.

The first test run to Lock 4 is under way. Mark Oser leads mules Jack and Jenny.

Paul Baird and Dave Richey, on bow deck, watch Walter Smith untangle the towline at the Lock 4 turnaround. Clyde Gainey stands at the gunwale, Gale Hartel watches from the roof, and Dick Mohler waits at extreme right.

Dick Beck paints the *St. Helena II*'s name on her transom.

The completed boat sign at the dock.

The *St. Helena III*. Photo courtesy of Canal Fulton Heritage Society.

Dick Mohler does janitor work in the partially framed middle cabin. The catwalk above him connects cabins.

From left: Roy Preece Sr., Paul Baird, and Walter Smith fabricate iron fittings for the boat.

Throughout the winter, to fabricate the many custom iron boat fittings, including deadeyes, rudder fittings, bow stem fittings, mooring rings, handrails, lantern mounts, and pike pole hooks, Roy Preece Sr., Paul Baird, Walter Smith, Paul Marks, and Pete Neidert worked in a makeshift blacksmith shop in Pete's tractor-service shop. At the same time, patterns for the huge capstan castings were prepared at the Hoover Company pattern

Al Simpson gets acquainted with "Red," one of the two mules purchased at Ralph Gore's farm. The other mule was "Itchy."

shop; these were later cast in iron at the United Engineering and Foundry Company. All hardware items were exact full-size replicas of those on the McLaughlin *St. Helena* boat model.

A historically correct power system was of course needed, and a March 22 visit by members of the Stark County Historical Society's Canal Branch Board to Ralph Gore's farm in Carroll County resulted in the purchase of two big mules, "Red" and "Itchy," who needed to be trained for canal boat towing duty—a bit more demanding than farm work, since they needed be taught to work as a team. A mule, the offspring of a male donkey and a female horse, is considered more patient, sure-footed, hardy, and long-lived than a horse, and is believed to be less obstinate, faster, and more intelligent than a donkey.

In April, the Canal Fulton boat acquired its formal name, following a public naming contest. A committee reviewed 150 submitted names. The winner, the *St. Helena II of Canal Fulton,* was the most appropriate name possible, recognizing the new canal boat as a reincarnation of the original *St. Helena.* Donations totaling $50,000 had been received from the

Hoover, Timken, and Stark County foundations; these enabled the hiring of a strong young crew who would be trained to operate the boat. The initial crew hired were David Richey, eighteen, of Canton; Scott Lehman, eighteen, of Canal Fulton; and, as captain, Randall Lemke, twenty-one, of Canton.

As plans were underway for the *St. Helena II*'s inaugural season, a significant national event occurred, one that would have a long-term impact on canal restoration. April 22, 1970, the first Earth Day, marked the beginning of the major modern environmental movement in the United States. Approximately 20 million Americans participated, with a goal of a healthy, clean environment. It was clear that people throughout the country were becoming receptive to environmental concerns. This would create much broader national support for cleaning and restoring the canals, through grassroots and private initiatives as well as federal grants and legislation.

On April 25, a number of trees along the canal side of the towpath that would have interfered with the *St. Helena II*'s towline were removed, and replacements planted on the riverside of the towpath. Dick Mohler completed

Clyde Gainey and Mitch Gantz take a break from construction.

cabin hatch covers and authentic cabin interiors with ladders, wainscoting, window frames, bunks, mule feed bins, and storage cabinets. The cabins were enclosed in redwood paneling, and all exterior surfaces were given a primer coat of brilliant white undercoat paint. Roofs, sliding hatch covers, and catwalks were covered with canvas to provide non-slip surfaces for crew, as well as protection from the weather. The trial runs were just over a month away. This final round of last-minute construction called for the usual army of volunteers, which always showed up when needed, ranging from my nine-year old son, Mitch, to seventy-five-year-old Clyde Gainey.

Dedicated as they were to the task at hand, volunteers were still very much aware of the technological triumphs as well as the terrible tragedies on the national scene that occurred during the construction years. With the shocking March Mai Lai massacre still reverberating in the news, the first two weeks of May 1970 were among the most traumatic of the anti-Vietnam War demonstration period. On May 1, President Nixon ordered U.S. forces to cross into neutral Cambodia, threatening to widen the war, and triggering violent antiwar protests across the country. On May 4, on the campus of Kent State University, only twenty-five miles from Canal Fulton, National Guardsmen shot and killed four student demonstrators. On May 9, a hundred thousand people gathered in Washington, D.C., to protest the war, and on May 14, Mississippi law enforcement officers killed two student demonstrators at Jackson State University. These traumatic events seemed almost unreal from the perspective of quiet, patriotic, and industrious Canal Fulton.

Several major tasks other than boat construction remained, which, if not accomplished, would prevent boat operations. On May 18, the footbridge that crossed the canal south of the dock and connected the village to the village park was raised six feet, to allow boat and mules to pass underneath. It was still a low bridge hazard to crew standing on the roof of the boat, but at four miles per hour, there was plenty of warning time to duck. On

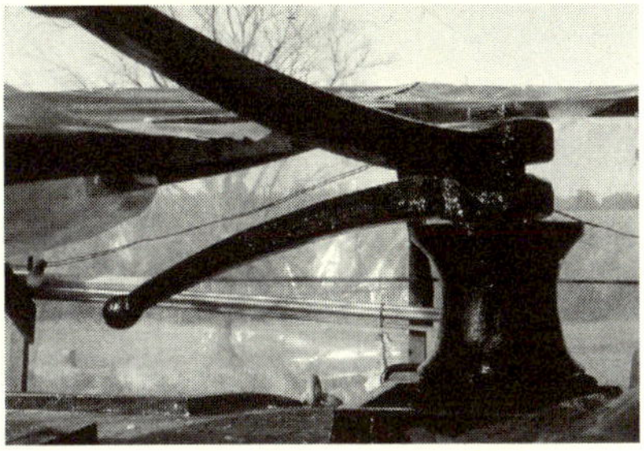

Above left: Stark County Engineers' maintenance department employees dredge the turnaround basin at the work site.
Above right: The capstan on top of the center cabin will raise and lower the gangplank.
Left: The steering deck, with the tiller attached to the rudder post.

May 23, a turnaround basin, a widening of the canal, was dredged exactly where the construction site had once been; this allowed the boat to be turned around after each trip before it took on new passengers, and the boat was moved into the new basin for completion. At the same time, a new docking facility and ticket house were erected on the canal in the village park, across from the turning basin.

By June 6, most of the plastic greenhouse was removed from the boat, and hardware was installed, including the three-hundred-pound cast iron capstan on the roof of the center cabin, and the steel support frames for the convertible canvas awnings over the passenger deck, needed to protect passengers in the event of rain or hot sun. Only two days remained before the first trial run, and Dick Mohler rushed to complete last-minute woodwork. His major task was the installation of the huge, heavy rudder, five by six feet and four inches thick. Scuba divers placed the nine-inch-diameter rudderpost into its underwater lower metal fitting. The top of the rudderpost on the steering deck was fitted with a wooden tiller for the steersman to control the rudder. Other last-minute details included the tiller deck rail, passenger benches in the fore and aft holds, and six bilge-pump wells in the

Last-minute painting the evening before the trial; note the remnants of the winter greenhouse are still attached.

Glenn Oser and his daughter, Carolyn, stand with mules "Red," "Itchy," "Jack," and "Jenny."

deck. A functional hand-operated bilge pump, just like on the McLaughlin model, was built to add authenticity, but a battery-operated modern pump was also installed to do the actual work. To hasten completion, painting by large volunteer crews continued well after dark on several nights.

The first test run occurred on June 8, with remnants of the greenhouse still attached to the boat. Local farmer and muleskinner Glenn Oser (1921–1993) housed and fed the mules on his farm, and brought them to the canal site with a trailer, labeled prominently the "Hee-HAW-ler." Glenn had two teams of mules ready for service, because they were trained in pairs. The original pairs were "Red" and "Itchy" and "Jack" and "Jenny," a reddish-colored team. Later, "Kitty" and "Katie" and "Milly" and "Molly," a black team, would replace the first teams, respectively. The mules were hitched in tandem and pulled the twenty-two-ton boat with a 150-foot towline.

The first test run gets under way. Randy Lemke stands at the tiller with Scott Lehman; Walter Smith is on the gunwale; Paul Baird and Gail Hartel are on the front cabin roof; and Dave Richey is on the bow deck.

The first test run to Lock 4 is under way. Mark Oser leads mules Jack and Jenny.

The test run to Lock 4 and back was intended to instruct the crew in handling the boat and to demonstrate how the mules behaved in actual operations. A number of key volunteers accompanied the crew. For most of us, this was a long-anticipated and thrilling event. Though we were familiar with every square inch of the boat, we had no idea what it would be like to actually be under way. The *St. Helena II* was no longer a mere construction; she became a living thing before our eyes.

We marveled as she slid quietly away from the village, and we were instantly transported back in time to the unique nineteenth-century experience of motion in a huge boat without noise, through the silent and scenic rural countryside, on a narrow, shining road of still water. In this modern era of mechanical transportation, movement is inevitably accompanied by a constant hum, rumble, clatter, splash, or swoosh, always with constant vibration. A canal boat is different. Total silence. Like the hundreds of thousands of tourists that would follow us, we drank in the overwhelming sense of ponderous but deliberate movement, realistic historical surroundings, and quiet peacefulness. It was a genuine nineteenth-century feeling. We could not help but marvel that our personal experience on an authentic operational canal boat was the first of its kind to occur in several generations.

But we had work to do. One of the key problems to be resolved was where to attach the towline to the boat. If the line were attached too far forward, the boat would be pulled into the towpath bank; too far aft, the boat would tend to veer toward the opposite bank. By trial and error, we discovered quickly that with the towline attached to the iron deadeye on the towpath side of the forward cabin, essentially at the boat's center of gravity, the boat was equally balanced to the angle of the towline, and very little effort

Paul Baird and Dave Richey, on the bow deck, watch Walter Smith untangle the towline at the Lock 4 turnaround. Clyde Gainey stands at the gunwale, Gale Hartel watches from the roof, and Dick Mohler waits at extreme right.

was required by the steersman to proceed directly down the center of the canal. The deadeye, incidentally, was positioned exactly as it had been on the McLaughlin model, proof of the model's functional authenticity.

The most complicated maneuver was turning the boat around in the basin at Lock 4. As the mules stopped towing, the captain had to steer the still-moving boat in a large counterclockwise circle. Boatmen had to grab the eight-foot pike poles and use them to keep the boat on course and from hitting the bank. One boatman had to release and shift the towline to the opposite side of the cabin and reattach it for the return trip. This was no easy maneuver with a twenty-two-ton boat in motion, but the crew handled it well, in spite of a towline tangle. On the final leg back to town, the mules were reluctant to go beneath the raised footbridge, but did so after some repeated urging by Glenn Oser. Later, the towpath under the bridge was widened and lowered, which gave the mules more space. In general, everyone considered the first test run successful.

The second test run was on June 12, this time without the greenhouse remnants. Final coats of white paint had been applied on most of the cabin and hull exteriors, with gray roofs, gangplank, decks, steps, and cabin siding; laurel-green shutters and passenger benches; black rub rails; walnut stain on all interior framing timbers; gold and cream in cabin interiors; and black on all metal fittings. The boat finally took on the appearance of a finished historical boat.

The heavy gangplank for passengers had been installed and attached by a rope to the capstan so the crew members could practice quickly raising and lowering it for passenger boarding and disembarking. They had to practice mooring the boat to the dock pilings with ropes after the second turn around in the construction site basin to discharge passengers. During the second run, several tree limbs along the canal were pruned to avoid striking boatmen standing on top of the cabins or catwalks. It was also determined that a complete trip took forty-five minutes. After these initial

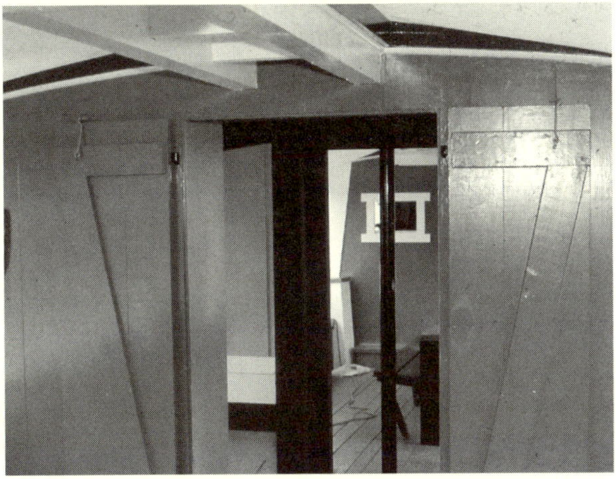

The painting completed, this view is through the center cabin toward the aft cabin.

The second test run. Crew members Dave Richey and Scott Lehman (hidden behind Dick Mohler) work the capstan to raise the gangplank while Randy Lemke (bottom) and Dick Mohler watch, and Gail Hartel (top left) communicates ship-to-shore.

trials, earlier plans to take the boat through the lock were abandoned, as that procedure would have taken an additional hour and required additional personnel. This seemed impractical, given the anticipated demand by so many passengers and the limited time during peak-attendance days to accommodate them. The board of managers concluded forty-five minutes to be the ideal trip time, so that there could be a scheduled ride every

Dick Beck paints the *St. Helena II*'s name on her transom.

The finished aft cabin interior.

Volunteer crew members wait for the press run: from left, senior volunteers Walter Smith, Paul Baird, and Clyde Gainey. Crew members Dave Richey and Scott Lehman wait on top of the boat.

Jim Guest welcomes the press with Junior League hostesses: Mrs. Vicki Conley, Mrs. Karen Williams, and Miss Sheila Markley (partly hidden).

hour, on the hour. This was a successful decision from an operational viewpoint. The board also decided that special group trips through the lock could be planned later; in fact, they would be run in the early 1980s until feeder problems arose.

June 17 was to be the all-important press run, when reporters from around the state were to ride and to write about the trip, the boat, and its formal dedication scheduled for July 11. The *St. Helena II* had to be at her cosmetic best for press photographers, and she was. On June 16, a professional sign painter had applied the bold name *St. Helena II* on her transom, in the same decorative graphic format as it had been on the original boat and model. Below was printed "of Canal Fulton" and at the bottom, "The Stark County Historical Society." Other finishing touches included a potbellied stove in the forward cabin, and a cook stove, antique pots, and bedding in the aft cabin. Passengers could view these furnished cabins through windows from the passenger deck.

During the press run, captain and crew were dressed in appropriate nineteenth- century working attire, and Canal Fulton Junior League hostesses were decked out in period costumes as they welcomed reporters and photographers to the dock for departure. Jim Guest served as the announcer of the event. Cameras snapped and notebooks were brandished. The forty-five-minute ride went off without a hitch, and the press spread the word very effectively. All the following week, comprehensive articles and photos describing the *St. Helena II,* the story of her development, and announcements of her upcoming dedication appeared in the newspapers of Canton, Canal Fulton, Alliance, Akron, and Cleveland. It was the *St. Helena II*'s first introduction to the broader public!

Volunteers and the crew made two final trial runs on June 27 to rehearse and refine the complex loading, turnaround, and docking operations. The Stark County Historical Society had erected a sign in the village park publicizing the "Historic Ohio Erie Canal Boat Ride aboard the authentic canal freighter *St. Helena II* of Canal Fulton." At the ticket office, the schedule of boat trips was posted: "Tuesday through Friday: 1 P.M. and 2:30 P.M.—Saturday: 1 P.M., 2:30, and 4 P.M.—Sunday: 1, 2, 3, and 4 P.M. Ticket office open 12 noon to 5 P.M. (closed Mondays)." Adult tickets were $1.50, and children under eight paid 75 cents. The boat capacity was sixty.

The completed boat sign at the dock.

St. Helena II, ready for her formal debut.

After three years of dedicated work, volunteers could finally heave a collective sigh of relief and enjoy a deep sense of accomplishment and personal pride. We had built a perfect replica of an actual nineteenth-century canal boat, the first in the nation, not with present-day materials but with natural oak trees cut and sawn into wooden planks, and caulked in the same manner as had our forefathers. Our physical accomplishment was impressive. The *St. Helena II* weighed twenty-two tons empty, was sixty feet long on the keel, and sixty-five feet, seven inches from stem to stern. She was thirteen feet wide, about eight feet high, and drew only eighteen inches of water afloat.

It was estimated that the total cost of purchased materials and the thousand hours of paid labor was approximately $50,000. If one added the fifteen thousand hours of volunteer labor, at the 1970 minimum hourly wage of $1.60, the total would have been about $75,000. The value of donated materials could have added another $25,000, to make a total of about $100,000, roughly three times the average cost of a new home at that time. At 2010 rates, this would be approximately half a million dollars, still a tremendous bargain.

In July 1970, the *St. Helena II* belonged to the many citizens of Stark County, at no cost. She was ready for her formal debut, and for the beginning of her long career of public education and recreation on the Ohio & Erie Canal, where her nineteenth-century namesake had been born. It would be a major Stark County public event, remembered for years.

CHAPTER 8

Dedication

Canal Fulton's fifth annual Olde Canal Days Festival on July 10, 11, and 12, 1970, attracted ninety thousand people in fine weather. It was the village's most gala festival to date, preceded by promotional articles in the *Akron Beacon Journal* (June 30), *Cleveland Plain Dealer* (July 1), the *Barberton Herald* (July 2), the *Canton Repository* (July 2, 5, 8 and 10), the *Alliance Review* (July 9), the summer issue of the *Canawler,* and the July 8 issue of the Canal Fulton newspaper, the *Signal,* which featured the full upper half of the front page with a color photo of the *St. Helena II,* as well as a detailed agenda and description of the planned weekend blockbuster events, including the crowning of a Canal Days Queen, a fire department parade, a water fight between Clinton and Canal Fulton volunteer fire departments, Up with People singers from Massillon, the (Canton Football) Hall of Fame Chorus, the Act IV barbershop quartet, amusement rides from the Ohio State Fair, an organ grinder with a trained monkey, a majorette unit on roller skates, tours of historic homes, a spectacular exhibit on Ohio's natural resources from the Wonderful World of Ohio, Komas the Hindu Fakir, a float parade on the canal, fireworks, a grand parade with high school bands and clowns, an inner tube regatta, the Sky Hawks sky diving team, the Within Six Months rock band for the Pitkin Ford parking lot dance, the Rotary Club's annual chicken barbeque, and, of course, the christening and the first public rides on the *St. Helena II,*

State senator Ralph Regula gives the *St. Helena II*'s dedication address, on the guest platform at the dedication ceremony. Seated, from left, are Canal Days Queen Carla Remark; Viona E. Findley, widow of Ted Findley, the late president of the Canal Society of Ohio; myself and my wife, Lorraine; Ed Harriman; James Dillow Robinson, of Cleveland, a founding member of the Canal Society of Ohio; Dr. R. K. Ramsayer, president of the Stark County Historical Society; Canal Fulton mayor J. W. Richardson; Mary Muhlhauser (1893–1980), key financial contributor; Gervis S. Brady, executive director of the Stark County Historical Society; Jim Guest of the Louisville Sportsman's Club, event chairperson; and Al Simpson of the *Canton Repository.*

not necessarily in that order. Clyde Gainey opened his Old Canal Days Museum free of charge for all visitors, in celebration of his seventy-fifth birthday. Commemorative dinner plates showing the various steps in building the *St. Helena II* were being sold.

More than any of the preceding four festivals, this one celebrated not only the canal days of the past but the future of this small town, which was now empowered by its pride and joy, the *St. Helena II*—pride because of the large number of residents who worked tirelessly for over three years to build her, and joy because of what she promised to bring this small town in terms of business, recognition, and employment. It was truly a memorable event!

The public dedication and christening of the *St. Helena II* took place at 10:00 A.M. on Saturday, July 11, in the village park boat landing, with Jim Guest as general chairman and Ed Harriman as Canal Branch Board chairperson and as master of ceremonies. Ed Harriman gave the invocation, and J. W. Richardson, mayor of Canal Fulton, welcomed attendees. Ed commended the many volunteers who had worked on the boat during the four years it had taken to design and build it and introduced the honored guests on the ceremonial platform.

After Ramsayer made a few remarks, Robinson (1899–1980), perhaps the last living Ohioan who had actually worked on the canal, read his original canal poem, "Canal Nostalgia," written for the occasion and dedicated to the settlers and pioneers of Ohio. (See appendix A for the complete poem.)[1]

From age six to seventeen (1905–17), James Dillow Robinson lived on a state canal boat with his stepfather, Captain Charles B. Stebbins, of Independence, Ohio, who was the repair foreman of the canal between Cleveland and Peninsula, repairing locks, aqueducts, breaks in the banks, and cleaning out feeders. The family lived on the boat year-round. In 1912, Dillow joined his stepfather's crew and was thus on the canal during the infamous 1913 flood that shut down the canal. He continued working for the state on a section from Pinery Dam to the American Steel & Wire Company, where it served as a water source. He quit in 1917 to go into the grocery business with his stepfather but joined the army in 1918 to serve in World War I. In 1922, he returned to canal maintenance on the same stretch, working as an employee of the U.S. Steel Wire Plant until 1924. In about 1934, he inherited his stepfather's combined grocery store and post office in the old South Park railroad station in Independence. He operated the store, and his wife, Mabel, served as postmistress for the fifty-two families of South Park. He worked as supervisor of Meyer Dairy Products Company from 1935 until his retirement in 1965. Afterward, he wrote articles for the Canal Society of Ohio and was an honorary trustee of the society.[2]

The Honorable Ralph Regula, who took the initiative in acquiring canal lands for Stark County, and who became the driving force in the state for the conversion of canal lands into recreational highways and trailways, then gave the dedication address. He was obviously delighted with the major milestone in his vision represented by the *St. Helena II*: "What a great day! I am an eternal optimist but even I began wondering if this would ever get done." After briefly summarizing canal history, he continued: "Today we launch *St. Helena II,* the only completed authentic canal boat in the nation—a truly historic occasion for us and for Ohio. We add a jewel to Stark County's crown—one that can proudly take its place with the Stark Wilderness Center, the Stark Historical Society and the Canton Hall of Fame. Today we add a chapter to the Stark County Story."

Regula then recognized the individual contributions of many volunteers and the enthusiasm and leadership demonstrated by the community, the county, and the state. He concluded:

> Today we launch more than a canal boat. The *St. Helena II* is a symbol that the pioneer spirit of individuals can achieve the impossible—and that it can be achieved using the resources and energies of the community alone. The thing is, we stuck together and we finished it.
>
> The success of this project stands as a challenge to all of us—to continue to preserve the heritage of our past and to challenge us to great goals in the future. Borrowing a phrase from one of Stark County's well-known industries we launch a symbol of our proud past and look forward to an exciting future that will include continued enhancement of Stark's ribbon of wilderness, its twenty-three miles of canal lands.[3]

(See appendix B for the complete Regula address.)

The highlight of the ceremony was the christening of the *St. Helena II,* an honor shared by Viona Findley and my wife, Lorraine, who jointly swung the bottle of champagne, suspended from a pole by Dick Mohler on top of the forward cabin, down against the wooden hull. The bottle bounced

Above, from left: Lorraine Gantz and Viona Findley hurl the champagne bottle, while Jim Guest (partly hidden) and Ed Harriman watch. Copyright *Canton Repository;* used with permission.

Right: At last, Randy breaks the bottle with the twenty-third hit. Photo courtesy of John C. Harriman

off the sturdy oak as if it were a rubber wall but remained perfectly intact. The ladies tried a number of times with the same unsuccessful result. Only after boat captain Randy Lemke took the bottle by its neck, waded into the water up to his waist, and slammed it against the iron rubrail for the twenty-third time, did the bottle finally break, and the ceremony was complete. The canal banks on both sides for a hundred yards were lined with people who cheered loudly upon seeing the bottle break. The *St. Helena II* was open for business.

At 11:00 A.M., the *St. Helena II* took her dedication voyage down the canal, with dignitaries, speakers, Canal Board members, volunteer lead-

Boarding *St. Helena II* for the dedication voyage are, from left, Dave Richey (crew), state senator Ralph Regula, Scott Lehman (crew), Al Simpson, the author, Ed Harriman, and Gervis Brady. Photo courtesy of Canal Fulton Heritage Society.

ers, and their special guests aboard. Some of these were the people who envisioned the idea of a boat and provided the organization, inspiration, and enthusiasm to arouse the community to action.

The second ride, at noon, was for the key construction volunteers and their spouses only, with a box lunch served along the way, to recognize and honor their dedication and hard work that physically created the *St. Helena II* over the last three years. For the first time, the volunteers were able to relax and enjoy the boat they had built, with their bare hands and out of thin air. I had the feeling however, that many, while appreciating the gratitude shown them for their accomplishment, regretted that their work was done. They had truly enjoyed the challenge of contributing to a project so important to the community. The accomplishment of what many considered impossible was even more rewarding. Unfortunately, it was difficult to similarly honor all five hundred or so volunteers who had also contributed time, money, and talent to the *St. Helena II.*

Public rides began at 1:00 P.M. and continued throughout the weekend. The crew performed perfectly, and there were no problems with the boat or with the some four hundred paying passengers who patronized the boat during her seven trips that weekend. After Randy cast off, the *St. Helena II* moved silently south past the park and under the footbridge. During the ride, a recorded lecture was played as the boat was pulled quietly down the scenic canal. If you were a passenger, you would have heard this recording: "The Stark County Historical Society welcomes you aboard the *St. Helena II* of Canal Fulton. In the next forty-five minutes, we would like to take you two and a half miles and 140 years back into history, when the Ohio canals opened the state to commerce, industry, and settlers. Your next forty-five minutes will be spent on this original portion of the Ohio-Erie Canal, which was begun on July 4, 1825."[4] (See appendix C for the transcript of this recording.)

As the *St. Helena II* passed under the footbridge and the noise of the

Captain Randy casts off.

crowd faded into silence, all were impressed with the trip. It could not help but remind them of the pristine, silent wilderness that the canal had passed through during its peak years of 1825 to 1860, and they must have marveled that in 1970, the feeling of that historical time and place still could be recreated so perfectly. They watched the mules on the towpath as they plodded along, and felt the ponderous yet graceful movement of the boat, as if it were propelled by some quiet and unseen force. For a brief period, they were transported back in time and place, away from twentieth-century traffic, crowds, speed, and noise. The peculiar sensation of disembodiment is truly hard to describe. The recording then continued to relate details about the path of the canal through the state, naming the many towns familiar to Ohioans. It described the difficult construction of the canal, noted the general types, functions, and designs of early boats, and then itemized the many construction details of the *St. Helena II.*

The boat leaves town, under the footbridge to the park.

The *St. Helena II* leaves Canal Fulton for open country.

The *St. Helena II* approaches Lock 4.

The recording optimistically speculated about future plans, which at the time included taking the boat through Lock 4 and even traveling to Massillon or north of Canal Fulton to the feeder canal. In fact, some charter trips through Lock 4 eventually occurred in the early 1980s, but regular operations remained only from the Canal Fulton village dock to Lock 4 and back. The recording also mentioned the Stark County Historical Society's plans to build a museum at Lock 4, but this never materialized. The recording, however, did correctly anticipate the McLaughlin drydock restoration, which would be accomplished magnificently in 1976. Over the years, costumed docents would replace this initial recorded lecture by narrating canal history and providing information in nineteenth-century style.

The turnaround at Lock 4 was the halfway point; it enabled the crew members to demonstrate their acquired skill in turning the boat smoothly and efficiently, with little loss of headway. As this was occurring, the recording described Lock 4 and its functional details of operation. The return trip to Canal Fulton allowed passengers to enjoy the quiet and restfulness of the natural environment. Finally, passengers were returned to the noise and activities of their departure point and were treated to another skillful boat turnaround before docking, disembarking, and watching passengers boarding for the next trip.

During and after the festival, coverage of the christening of the *St Helena II* was featured in the *Canton Repository* (July 11, 12, and 13), the *Akron Beacon Journal* (July 12), and the *North Canton Sun* (July 15). Many newspapers and magazines around the country published feature articles describing the arrival and construction story of the *St. Helena II*, including the August 9 *Columbus Dispatch* Sunday magazine, the Timken Company's national magazine, the home section of the August 29 *Akron Beacon Journal*, the August *Ohio Edisonian* magazine, and the September 13 *Toledo Blade* Sunday magazine.

Glenn Oser halts the mules at the Lock 4 turnaround.

The *St. Helena II* makes her return trip to Canal Fulton.

During July operations, the *St. Helena II* carried 4,129 passengers and in August nearly 6,000. By the end of the season, on October 26, 15,554 passengers had enjoyed rides. As winter approached, she was wrapped in a cocoon of plastic for protection and moored at her dock in Canal

Scott Lehman and Dave Richey make the final push into dock.

Fulton Village Park. In the 1971 season, with a new captain—Terry Finefrock, the husband of Bonnie Oser, Glenn's daughter—she carried 29,385 passengers, almost double the 1970 number. Thus, in just one and a half seasons, assuming an equal number of adults and children paid fares of $1.50 and 75 cents respectively, the *St. Helena II* generated approximately $50,000, her original cost in purchased construction materials and paid labor. This demonstrated to canal communities everywhere that replica boats were well worth their initial financial investment.

Meanwhile, the renovation of the entire Ohio & Erie Canal moved forward. The National Trails System Act of October 1968 had promoted the preservation and public access to hiking trails for scenic, historic, and recreational purposes. The towpath in Stark County became part of the Buckeye Trail on October 10, 1970, when the five-hundred-mile hiking trail from Cleveland to Cincinnati was dedicated by none other than state senator Ralph Regula along with senator Harry L. Armstrong, who had co-sponsored a resolution establishing the trail as Ohio's official hiking route. Today, the trail stretches for thirteen hundred miles in twenty-four sections and touches forty of Ohio's eighty-eight counties. There are 66,438 miles of such trails nationally.

The *St. Helena II* of Canal Fulton was just the beginning of a national movement to build operational replicas of nineteenth-century canal boats on restored canals. Soon, it was apparent that such boats could provide not only recreation for the public but an attraction to historical sites, increased appreciation of our canal heritage through a more realistic presentation, and a profitable resource for parks and communities. Other canal communities in Ohio, and later in many states throughout the East, would follow Canal Fulton's bold lead.

CHAPTER 9

Canal Boats USA

During the 1950s and 1960s, several canal societies had promoted the history of canals and the preservation of canal artifacts such as locks and aqueducts in their geographic region: The Chesapeake & Ohio Canal Association in Maryland (founded in 1954), the Canal Society of New York (founded in 1956), the Canal Society of Ohio (founded in 1961), the Middlesex Canal Association in Massachusetts (founded in 1962), the Pennsylvania Canal Society (founded in 1966), the Delaware & Hudson Canal Historical Society in New York (founded in 1968), and the Canal Society of New Jersey (founded in 1969). Their motivation was to retain, record, recover, restore, and re-create the canal era. These societies published regular newsletters for their many dedicated members, who on field trips tracked down, photographed, and recorded surviving canal locations, conditions, remains, locks, artifacts, aqueducts, and sites. Many of these organizations had been keenly aware of the *St. Helena II* project and enthusiastically published status reports, as well as encouragement, to their local canal communities, some of which also caught canal boat fever. It is significant that the first three replica boats in the country were in Ohio.

The second replica boat launched was in Coshocton, Ohio, seventy miles south of Canal Fulton, down the O & E Canal, where "Mad" Marshall Jacobs (1909–1977), a well-known flagpole sitter and steeplejack (in 1946, he was married on top of a 180 foot flagpole, an event featured in *Life* magazine), had begun construction of a seventy-eight-foot canal boat in July 1968, with timbers from an old grain-storage building. By July 8, 1971, the mule-drawn *Monticello II* was completed and began trial runs on a two-mile section of the canal at Lake Park. Named after the first canal boat to arrive in Coshocton in 1830, she weighed twenty-five tons and as an open packet passenger boat could carry 120 passengers. She began public operations in September 1972 in the restored Roscoe Village in Coshocton.

Encouraged by the resurgence of interest in canals, the American Canal Society was founded on January 1, 1972, with members from the Canal Society of New York State, the Chesapeake & Ohio Canal Association, the Middlesex Canal Association, the Pennsylvania Canal Society, the Delaware & Hudson Canal Historical Society, the Canal Society of New Jersey, the Canal Society of Ohio, and the Canal Museum in Easton, Pennsylvania. With its cofounders, Thomas Frederick Swiftwater Hahn (1927–2007, its first president), William H. Shank (1915–2007), and William Trout III, this new national canal organization began publication of its newsletter, *American Canals,* in March 1972. Now, with a national canal newsletter, the success of the *St. Helena II* and her inspiring local story reached far and wide. Within just a few years, interest in canal boat replicas mushroomed into a national trend, resulting in a wave of canal boat constructions and creating a veritable national fleet of replica canal boats that continues today.

Spring 1973 saw the start of mule-drawn operations of the third Ohio boat, the *General Harrison,* a replica of an 1840 canal boat *Clint Hosher,* built from 1969 to 1971 for the Ohio Historical Society on a mile-long section of the Miami & Erie Canal at the Piqua Historical Area in Piqua, Ohio. The society opened the area to the public in 1972 at the John Johnson farm with a museum telling the story of the Eastern Woodland Indians of Ohio. A freight boat similar to the *St. Helena II,* the *General Harrison,* was seventy-seven feet long by twelve feet wide, was towed by horses, and carried 120 passengers.

The replica canal boat fever spread beyond Ohio, including New York, the state that boasted the first major state canal system in the country. Its boat was the *Independence,* constructed from 1972 to 1973 in Rome, New York, on the Erie Canal, by the Historic Rome Development Authority, using plans by Robert E. Hager, with construction supervised by Curry Bartlett. The *Independence* began horse-drawn operations at Erie Canal Village near Medina in July 1973 and was seventy feet long by twelve feet wide, carrying 125 passengers. She was later renamed *The Chief Engineer of Rome,* for the first boat to travel on the original Erie Canal.

The next state to catch replica canal fever was Pennsylvania. In fall 1974, the *A. Emerson,* a half-scale replica of boat #1107, named and modeled after the last coal boat to leave Honesdale, Pennsylvania, in 1898, began mule-drawn operations at Lonesome Lock 34, four miles from Honesdale, on a restored section of the Delaware & Hudson Canal, a 108-mile canal that carried coal from Honesdale to the Hudson River. She was made of redwood, weighed thirteen tons, was fifty feet long, and carried fifty passengers. In 1977, the Canal Campsites Company sold her to Hugh Moore Historical Park and the National Canal Museum in Easton, Pennsylvania, for $10,000. Under her new name, *Josiah White*—for the founder of the Lehigh Coal and Navigation Company, which originally built the canal—she started mule-drawn operations in 1978 on a section of the Lehigh Canal with three operating locks. In fifteen years, she carried 180,000 passengers and earned $500,000 for the historical park and museum. Also in fall 1974, a mule-drawn barge constructed by David and John Knox began operations at Lewistown, Pennsylvania, on a quarter-mile section of the Pennsylvania Main Line Canal, which the brothers rewatered from the Juniata River. In 1975, the barge was named the *Juniata* and improved with a roof to resemble a canal boat.

In 1976, during the nation's Bicentennial, the whole country was reminded of its historical heritage, and public interest and support increased for the restoration of historical buildings and structures, including those along the canals. Tourism to historical museums and sites increased dramatically. That year, the National Park Service (NPS) launched a new mule-drawn *Canal Clipper III,* a wire-reinforced concrete hull simulating a C & O freight boat of the nineteenth century, in Georgetown on the Chesapeake & Ohio Canal. In 1979, the seventy-by-eleven-foot boat was relocated to the C & O Canal Museum at Great Falls Tavern in Potomac, Maryland, and was operated there by the NPS, including a half-mile trip through Lock 20 for its sixty passengers. It operated until 2003, when structural failure caused discontinuance.

In Massachusetts, the *Colonel Baldwin,* a horse-drawn replica of the 1803 Middlesex canal packet boat of the same name, began operations in 1977 on a restored stretch of the Middlesex Canal in Woburn, starting at the Baldwin Mansion built in 1661 (today a restaurant, The Baldwin) by Henry Baldwin, English emigrant and one of the town's founders in 1640. The *Colonel Baldwin* was forty feet long, nine feet wide, and carried forty passengers. The twenty-two-mile Middlesex Canal was one of the earliest in the nation.

Not all reconstructions were operated on canal waters. In 1975, a Naval Reserve Seabee unit constructed a land-based ninety-three-foot replica of a C & O canal boat, the *Cumberland,* at Lock # 75, five miles south of Cumberland, Maryland. The boat, completed in 1979, was based on the 1939 lines of C & O freight boat # 57 drawn by the NPS from actual boat remains in Hancock, Maryland.

In 1980, the *Valley Belle,* the thirty-five-passenger pontoon boat operating since 1965, was replaced with the seventy-five-foot long *Ben Franklin II,* a forty-eight-foot-long horse-drawn packet boat replica that weighed fifteen tons and had a capacity of seventy. Under the direction of Amos B. Schwartz, Indiana state engineers prepared specifications used for construction. The boat operated in Metamora, Franklin County, Indiana, at the Whitewater Canal State Historic Site until 1989.

In 1981, the Tri-Boro Sportsman's Club salvaged and restored an intact Lehigh Canal Boat (#249) from its fifty-year watery grave in a quarry near Lock 33. It was one of many scuttled there in 1931 by the Lehigh Canal & Navigation Company, getting rid of its deteriorating fleet after its operations ceased due to bankruptcy.

In September 1982, a new eighty-seven-foot fiberglass hull replica, *Georgetown,* also based on the 1939 NPS lines of C & O freight boat # 57, was built by Fiber Glass Fabricators, of Pasgagoula, Mississippi, with construction plans prepared by William G. Preston of Marine Power, Inc., in Gulf Breeze, Florida, and began half-mile (occasionally three-mile) operations through Lock 4 at the NPS Georgetown Visitors Center, carrying seventy to ninety passengers. Thomas Hahn, president of the American Canal Society, and myself, designer of the *St. Helena II,* were both consultants on the design. The *Georgetown* was still operating in 2010.

In July 1984, the *Sandpiper,* a steel-hulled replica of an Ohio canal freight boat of the 1800s, began operations on the Maumee River past the old Swain Creek canal outlet to Promenade Park in Toledo's rejuvenated waterfront area. Diesel-powered, she was sixty-five feet long, weighed thirty-seven tons, and carried one hundred passengers.

In 1987, three Lockmaster canal boats, designed to resemble English-style narrow boats and built by Mid-Lakes Navigation of Skaneateles, New York, became available for rental on the Erie and Champlain Canals. They were steel-hulled and forty-one feet long. Another land-based Ohio canal boat was constructed in St. Marys, Ohio, called the *Belle of St. Marys.* She was built in 1989 on concrete and steel stationary supports in the short level of the Miami & Erie Canal in a park between Locks 12 and 13, in Troy, Ohio.

In 1992, a sixty-five-foot reconstruction, transportable for educational purposes, *Marguerite II,* was built by the Delphos Canal Commission based on remains found in Delphos, Ohio, and was located two blocks south of Delphos, in Troy, on the Miami & Erie Canal.

In New Hope, mule-drawn barges had been regularly operating since 1955 on the Delaware Canal in Bucks County, Pennsylvania. In 1974, the canal was placed on the National Register of Historic Places, and in 1976, it was designated a National Historic Landmark. From 1977 to 1996, concessionaire George Schweickhardt ran tour operations with four operating barges and provided as many as ten mules from his farm. In 1983, the Friends of the Delaware Canal (FODC) was formed as a watchdog group, and in 1989, the park was renamed the Delaware Canal State Park. New aluminum and plastic lumber barges *Molly Pitcher* (later renamed *Priscilla Jean Pitcher*) and *Myfanwy Jenkins* (later renamed *Isabelle Justice*) were built in 1996–97 and ran until 2004, when flood damage to the canal suspended operations, but they were planned to resume in the future.

In 1994, a mule-drawn replica of an Ohio freight canal boat, *The Volunteer,* began operations in a reopened segment of the Miami & Erie Canal at Providence Metropark in Grand Rapids, Ohio, twenty-five miles south of Toledo. Sixty feet long, she carried seventy-five passengers on a forty-five-minute trip through working Lock 44 beside the Isaac Ludwig Mill, an 1846 working water- and steam-powered saw and gristmill.

An authentic English narrow boat, which English boat builders constructed in 2000 for the Blackstone Valley Tourism Council, was put into operation at the Blackstone Valley Visitor Center, Pawtucket, Rhode Island, the same year. Forty feet long, she was named the *Samuel Slater,* to commemorate the English immigrant who came to Rhode Island in 1789 to found America's cotton textile industry by building, from memory, operating replicas of Richard Arkwright's 1769 cotton-spinning machine, a British top-secret technology at the time.

In 2004, the *Lois McClure* was launched at Burlington on Lake Champlain, Vermont. She is an authentic full-scale replica of an 1862 canal schooner with sail, constructed by the Lake Champlain Maritime Museum. Typical of such schooners that operated on the Northern (or Champlain) Canal, built in

1823 to connect Lake Champlain with the Hudson River, she is eighty-eight feet long and fourteen feet wide, designed to sail from distant lake ports via the St. Lawrence or Hudson River, then lower sails and enter the canal locks. Volunteers at the Lake Champlain Transportation Company's Burlington Shipyard built her, using plans by naval architect Ron A. Smith. In 2007, she took a two-month, thousand-mile Grand Canal Journey on the Erie Canal to commemorate the Erie Canalway National Heritage Corridor, established in 2000. In 2008, she traveled north to Quebec City, Canada, commemorating the city's 1608 founding by French explorer Samuel de Champlain, and then sailed on to Montreal.

The western-most mule-drawn canal boat debuted in 2008 in LaSalle, Illinois, on a watered section of the Illinois & Michigan Canal. A classic seventy-six-foot Ohio packet boat replica with seating on top of the long cabin, she carried eighty-one passengers and was called *The Volunteer.* In 1999 the Canal Corridor Association of Chicago had engaged me as a design consultant, based on my design of the *St. Helena II.* I designed her based on photos and descriptions of historical packets. Scarano Boat Building, of Albany, New York, constructed her at a cost just shy of $1 million. Also in 2008, a land-based replica of a freight boat, the *Walter C. Pratt,* was built at the Black River Canal Museum at Boonville, New York.

The most recent arrival to the national fleet of canal boats is the *Delphi,* launched in June 2009 on a watered section of the Wabash & Erie Canal in Delphi, Indiana, and built by Scarano Boat Building, of Albany, New York. The *Delphi,* a replica of a canal freight boat, takes visitors on a thirty-five-minute tour. Community efforts to develop the canal and park in Delphi began in 1971.

In 2010, another I & M canal boat replica, the *Rosalie,* was refurbished by volunteers in Ottawa, Illinois, and placed on concrete pillars at the restored historic tollhouse on the lateral canal between Columbus and LaSalle streets in Ottawa, just a block south of the I & M Canal. The boat had been constructed for a 2006 film, *Prairie Tides,* and is intended to become a tourist attraction.

The foregoing includes several dozen replica canal boats built in the forty years since the *St. Helena II,* from Massachusetts to Illinois, all celebrating and re-creating a historical, educational, and recreational environment of the canal era. A number of additional boats being operated were just as successful commercially, and although less than historically authentic in design, they still restore community pride in local heritage, provide visitors with historical education and recreation, and encourage local economic and environmental development. Canal museums are no longer merely static displays of historical artifacts, but realistic, interactive, and participative experiences. Canal boats are ideal realistic, interactive, and participative exhibits.

But after the first few years of operating replica canal boats, communities had to deal with the ongoing problems of maintenance. The earliest of these replica boats, including the *St. Helena II,* were built with wooden hulls, because that was historically accurate. But such dedicated authenticity has its disadvantages. In the nineteenth century, labor was cheap, and functioning drydocks were numerous. Boats were placed in drydocks annually, so rotting planks could be replaced and recaulked. In the twentieth century, labor rates and absence of drydocks made maintenance of wooden boats so cost-prohibitive that almost all commercial and private boats have steel, aluminum, fiberglass, or ferro-cement hulls, to lengthen their life in water. So, since 1985, most of the replica canal boats mentioned above have used more modern, durable hull materials.

Not so for the *St. Helena II* and other early boats. After seven years of operation with irregular annual maintenance because there was no drydock to get her out of the water for repairs, the *St. Helena II* was leaking badly. In 1976, boat operations were transferred, by mutual agreement, from the Stark County Historical Society to the Canal Fulton Heritage

Society, which had planned since 1972 to renovate the old McLaughlin drydock, abandoned since 1913. Only partial earthen berms remained, but by installing a new sluice gate in the towpath, by rebuilding berms, and by filling the depression with canal water, the *St. Helena II* could be moved into it, positioned over raised stanchions, and left high and dry by draining the water from the depression into the Tuscarawas River. Volunteers replanked, recaulked, and repainted the worst sections of her rotted hull in 1976, and she resumed operations in summer 1977. Each year this process was repeated. But over the next eleven years, it became ever more difficult to find enough volunteers to perform this essential and extensive reworking. By 1988, leakage became so incessant and severe that she had to be removed from service. After eighteen years and five hundred thousand passengers, averaging twenty-seven thousand per year, the *St. Helena II* finally gave out.

In Canal Fulton, Ohio, Gale Hartel, one of the original volunteers of 1967–70, and Brooks Boat Design, designed a replacement for the *St. Helena II,* fittingly named the *St. Helena III,* with a ferro-cement hull, replicating the original boat design. The Canal Fulton Heritage Society erected a building over the restored drydock to permit year-round work. The construction process was quite different than that for a wooden hull. An inch or so of ferro-cement was applied over a hull shape made of thin wood slats and chicken wire. When dry, the cement hull weighed about the same as a wooden hull, but it required only minor annual maintenance. The *St. Helena III* began operations in 1992, continuing the Canal Fulton canal boat tradition. The Heritage Society still owned the boat, but the City of Canal Fulton operated and maintained it and the canal.

The *St. Helena II,* by then resting on the bottom of the four-foot-deep canal, was removed by crane in 1994 and transported to the nearby Canal Museum so volunteer citizens of Canal Fulton could completely restore her as a stationary outdoor exhibit; the task initiated in 1998 was labori-

The *St. Helena III.* Photo courtesy of Canal Fulton Heritage Society.

ously completed in 2000. There she rests now, by Clyde Gainey's Old Canal Days Museum and the Heritage House, in what is called St. Helena Heritage Park.

Similar maintenance problems plagued other early replicas. Indiana's *Ben Franklin II* was replaced in 1990 by *Ben Franklin III,* which had a fiberglass hull built by the Pasgagoula, Mississippi, Marine and Industrial Fiberglass Corporation. She was seventy-five feet long and had a capacity of eighty. That same year, Coshocton, Ohio's, *Monticello II* was replaced with *Monticello III,* built by Marine Builders, of Utica, New York, following plans by Jonathan Sandvich of Planning Resources of Cleveland. *Monticello III* had a stainless steel hull, was seventy-four feet long, fourteen and a half feet wide, weighed twenty-five tons, and handled 125 passengers.

In 1993 the original *Josiah White* on the Lehigh Canal in Hugh Moore Park, Easton, Pennsylvania, was replaced by the forty-ton, double-decker *Josiah White II*. She had a steel hull built by BethShip of Sparrows Point, Maryland, a canvas cover over the top deck, and a wooden superstructure built by Marcus Brandt Restorations to accommodate 150 passengers and take them directly to the museum. The cost was $775,000. In 1996, a new National Canal Museum was completed in downtown Easton, replacing the older structure.

In 1991, flood damage to the canal had halted operations of the *General Harrison* in Piqua, Ohio, and this boat was replaced in 2001 by the *General Harrison of Piqua,* a seventy-foot mixed-cargo boat, with a design similar to the original, but with an aluminum hull, built by Scarano Boat Building, of Albany, New York. She carries seventy-five passengers on a forty-minute, one-mile ride.

The *Canal Clipper III* at Great Falls Park in Maryland, because of structural failure, did not operate from 2003 to 2006, when it was replaced by the fifty-eight-by-twelve-foot aluminum-hulled *Charles F. Mercer,* a two-deck packet with passenger capacity of seventy-five. Built using cedar and Douglas fir by Scarano Boat Building, of Albany, New York, its $537,000 construction cost came from private donations, the State of Maryland, Montgomery County, and, most inspiringly, $3,000 solicited by the third-grade students at Seven Locks Elementary School, who stood at each lock and sold $10 and $20 passports offering free attendance to the C & O National Historical Park in Potomac, Maryland. Pulled by mules Molly and Ada for a half of a mile through Lock 4, the *Charles F. Mercer* was named after the first president of the C & O Canal Company, which reconstructed the original Potowmack Canal from Georgetown to Cumberland, Maryland, starting in 1828.

In addition to these mule-drawn replica canal boats, there are canal boat rides operating on the Erie Canal in a number of New York locations: Herkimer, Schuylerville, Waterford, Fairport, Dutchman's Landing, Lockport, Spencerport, Camillus, Macadon, and Rochester. Other rides operate in Richmond, Virginia, on the James River and Kanawha Canal, and in Augusta, Georgia, on the Augusta Canal. Detailed current information on all U.S. canal boat rides can be obtained on the American Canal Society website, www.americancanals.org.

The *St. Helena II* and subsequent replica canal boats were the first important physical, visible symbols of canal restoration. They attracted the public to old canal sites, publicized the importance of canal restoration, and encouraged local economic and recreational development. But they were only the sparks that ignited a firestorm.

The boats inspired a wave of canal lands development, restoration, and environmental reuse across the nation, which enriched public appreciation, increased recreational utilization of this national heritage, and ensured its protection far into the future. The next chapter describes some of these canal land developments in Ohio, which coincided with the state's canal boat replica constructions.

CHAPTER 10

Ohio & Erie Canalway

When he dedicated the *St. Helena II* in 1970, Ohio state senator Ralph Regula said: "The success of this project stands as a challenge to all of us—to continue to preserve the heritage of our past and to challenge us to great goals in the future. Borrowing a phrase from one of Stark County's well-known industries we launch a symbol of our proud past and look forward to an exciting future that will include continued enhancement of Stark's ribbon of wilderness, its 23 miles of canal lands."[1]

An exciting future indeed did follow. Senator Regula had not only inspired the county's efforts to assume responsibility for its twenty-three canal miles in 1964 and encouraged the community to build the first authentic replica canal boat in the nation but had from the start envisioned even more: a 309-mile hiking trail along the towpath from Cleveland to the Ohio River. For his entire career, he would pursue and support the challenge of this dream for the State of Ohio.

In 1972, a little more than two years after *St. Helena II* was launched, Ralph was elected a Republican member of the U.S. House of Representatives from Ohio's 16th District; he served eighteen consecutive terms in that role, until his retirement in January 2009, which made him the second-longest-serving Republican member of the House. He diligently and effectively used his long career and political influence to encourage a coalition of private and public organizations in Ohio that worked together to restore and develop the Ohio & Erie Canal lands in northeast Ohio for use by all Ohioans. It is his most enduring legacy.

When Representative Regula was first elected to the U.S. Congress, only a few small historical outposts of the northeast Ohio & Erie Canal were developing—a half-dozen jewels of isolated historical sites on a several hundred-mile-long canal silver ribbon. In the southern, mostly rural Tuscarawas Valley, there were three.

Canal Fulton didn't stop working after building its trendsetting canal boat. In 1972, I initiated a community project to restore the inherent architectural charm of the canal era in the village downtown area. I formed a design committee and invited the community to a forum for a presentation by Medina, Ohio, commercial artist Elmer Zarney (1922–2007), who since 1967 had spearheaded a facelift rejuvenation of downtown Medina, fifteen miles west of Akron. Known as the Medina Miracle, the project would be completed in 1973, when the public square would be placed on the National Register of Historic Places. Canal Fulton followed the Medina strategy. One by one, I prepared artist renderings to show property owners how their buildings might look with new paint and minor modifications of signage to present a more historical look, and, one by one, my suggestions were followed. Ann McLaughlin (now Henman), head of the Canal Fulton Heritage Society, continued these efforts in the 1980s. The result was a more historical look of the town, and today, 25 percent of the city's downtown structures are on the National Register of Historic Places.

Roscoe Village in Coshocton (Roscoe was a separate village across the Muskingum River from Coshocton in canal days) is a restored canal town seventy miles south of Canal Fulton on the Ohio & Erie Canal operated by the Roscoe Village Foundation. Edward and Francis Montgomery of Coshocton were the visionaries who made Roscoe Village a successful enterprise, by founding Roscoe Restoration in 1967. The Montgomerys wanted to restore the Whitewoman Street area on the west side of town into a working museum recalling the nineteenth century. In 1968, the community formed a support group, the Roscoe Village Society. By 1969, the group had converted a warehouse on the village square into a restaurant. Nearby buildings on the square, including a tollhouse and the Roscoe Building, were restored. During this same period, the Coshocton Chamber of Commerce developed a new recreational facility along the canal. By 1972 the replica canal boat *Monticello II* became operational, a perfect addition to Roscoe Village, which was added to the National Register of Historic Places in 1973. By 2005, thirty-five buildings in Roscoe Village had been restored.

Forty miles north of Coshocton on the canal was the Zoar Village State Memorial, in Waynesburg, Ohio, a restored 1817 settlement of German Separatists who helped build the canal. Although the original settlement was dissolved in 1898, villagers in 1930 formed the Zoar Historical Society and re-created the Zoarites' historical community garden. In 1936, during the Great Depression, the Zoar Foundation created a museum in Number One House (built in 1835) and worked with the State of Ohio to restore buildings. The State purchased the Zoar Garden, the Gardener's House, and Number One House in 1941 and the Tinshop in 1942. The Bimeler Museum was acquired in 1943. Development stalled due to lack of funds until 1968, when the state appropriated $300,000 for improvements. In 1970, the Zoar Store (built in 1833) was acquired, and it was restored by 1980. In the early 1970s, the State of Ohio began using costumed docents in the Tin Shop, Wagon Shop, and Blacksmith Shop. In 1995, the Dairy and the Kitchen/Magazine complex were restored, and in 1996 the State bought the Zoar Hotel (built in 1833) in the hopes of turning it into a visitor center.[2]

In the northern, most populated area of the Cuyahoga Valley, between Cleveland and Akron, the only historical preservation groups in 1972 were the Peninsula Valley Heritage Association, and the Hudson Historical Association, both formed about 1962 to restore and preserve the towns' historical New England village appearances. Residents Lily and Franklin Fleder and architect–interior designer Robert L. Hunker were instrumental in initiating and driving the Peninsula efforts, and in 1974, Peninsula became the fourth Ohio town added to the National Register of Historic Places by the U.S. Department of the Interior since its inception in 1966 (Zoar was added in 1969; Roscoe Village and Medina in 1973). In 1972, the Peninsula Valley Heritage Association became known as the Cuyahoga Valley Association (CVA), which formed the Cuyahoga Valley Park Federation in 1974 to link eighty local groups that wanted a federal park in the Cuyahoga Valley.

John F. Seiberling (1918–2008), Ohio's 14th District U.S. Democratic congressman (1971–1987) and chair of the Tri-County Regional Planning Commission, submitted a bill that created the thirty-three-thousand-acre Cuyahoga Valley National Recreation Area (CVNRA) as an act of Congress in 1974. It was at the time, however, a national park in name only. In 1975, Al Simpson, of the *Canton Repository,* went to Washington as Congressman Regula's press secretary and joined in the canal preservation crusade. Over the next thirty-four years, Regula guided over $200 million toward the development of the park. In 2000 it was renamed the Cuyahoga Valley National Park (CVNP). Development, under the leadership of park superintendent John Debo (1988–2000) and his predecessors, included a number of facilities, including the National Park Service (NPS) headquarters (1986), a Canal Visitors Center at Lock 38 (1989), Hines Hill Conference Center (1993), Cuyahoga Valley Environmental Education Center (1994), and Boston Store Visitor Center (1996), which featured part of a canal boat and tools related to boatbuilding. The CVNP partnered with nonprofits such as the Cuyahoga Valley Scenic Railroad (CVSR), which had initiated rail tours from the Cleveland Zoo to Akron in 1972. In 1989, the CVSR entered into agreement with the NPS, servicing the park and recently extending its service south to Canton.

In 1984, when the NPS decided to restore twenty-two miles of towpath within the park, some envisioned the possibility that the towpath might become the thread that could link Cleveland to Akron. In that year, the North Cuyahoga Valley Corridor, Inc., (NCVCI) was founded; in 1992 it became the Ohio Canal Corridor (OCC). In April 1990, groundbreaking for the Towpath Trail was celebrated in Cuyahoga Valley National Park.

In the late 1980s, Ann McLaughlin Henman of Canal Fulton, formerly head of the Canal Fulton Heritage Society, developed what would become a blueprint for an O & E corridor as an academic argument for her communications degree as a student at the University of Akron. Ann McLaughlin Henman; Christine Freitag, president of the Cuyahoga Valley Association; Peg Bobel, executive director of the association; John Debo of the CVNP; John R. Daily, Akron Metro Parks secretary-director; John F. Seiberling, recently retired as Ohio's 14th district representative to the U.S. House of Representatives; Virginia Wojno Forney of the Cascade Locks Park Association; and other interested participants met at Canal Fulton Village Hall on September 14, 1989, and organized the Ohio & Erie Canal Corridor Coalition (OECCC). A steering committee met monthly in the basement of the Old Canal Days Museum in Canal Fulton. Later, when "Canalway" became the official name for the corridor, the name was changed to the Ohio & Erie Canalway Coalition (OECC).

In 1990, Congressman Regula obtained $175,000 in federal funds to develop a feasibility study: "A Route to Prosperity." By 1992, the Cuyahoga County Planning Commission and the nonprofit North Cuyahoga Valley Corridor, Inc., produced a plan, titled "North Cuyahoga Valley Corridor." In 1993, Regula submitted H.R. 3593 to Congress, which designated eighty-seven miles of the Ohio & Erie Canal as a National Heritage Corridor, extending from Cleveland to Zoar, the historical canal village where the North Country Trail, a national scenic trail, would meet the canal towpath trail.

At the time, there were only three other sites designated as national heritage corridors: The Illinois & Michigan Canal Heritage Corridor (1984, 120 miles) in Illinois; the Blackstone River Valley National Heritage Corridor (1986, 43 miles) in Massachusetts and Rhode Island; and the Delaware & Lehigh National Heritage Corridor (1988, 165 miles) in Pennsylvania. The Regula bill passed in the House but was blocked from coming to a vote in the Senate, because some feared infringement on private property owner rights, although this was later proven not to be the case.

While federal legislation stalled, several communities were developing plans to extend the corridor from both the north and the south to link with the Cuyahoga Valley National Park. The City of Akron and the Akron Metropolitan Park District were looking at a Cascade Valley Park plan to connect from the center of Akron, north to the national park via the O & E towpath trail. In 1976, Walter Sheppe, founder of the Portage Trail Group of the Sierra Club, had proposed a similar plan to restore the Mustill Store (built in 1850) and house (built in 1841), located adjacent to Lock 15, both of which were restored in 1994. In the 1830s, this area was the town of Cascade, located at the foot of the seven locks that climbed up to Akron. By 1979 Akron had developed a Cascade Valley Park Master Plan.

In October 1983 the City of Akron restored and dedicated Lock II Park in the center of town, as originally advocated by the Canal Society of Ohio in 1963, to commemorate the location of William Payne's 1873 boatyard and drydock where the first *State of Ohio* boat was built. The park features a full-size, three-dimensional, steel skeletal frame of *St. Helena II,* constructed of white steel tubing through which visitors can walk and read a commemorative plaque engraved with the actual boat plans.

Virginia Wojno Forney had been involved with Progress Though Preservation, a new historic preservation group in Akron. In 1989 she formed the Cascade Locks Park Association, which would after 1992 help the

Summit County Metro Parks create Cascade Locks Park as part of Akron's Cascade Valley Park; the park is still under development. In 2002 the group interpreted the site where the waterwheel-powered Schumacher Cascade Mill (built in 1840) had been located. The park would make accessible to hikers the famous half-mile Akron staircase of seven Cascade Locks. When downtown Akron buildings were demolished in the 1990s, a park was constructed at Lock 3 and is now the city's hub of entertainment with music, festivals, and special events.

In the north, in 1992, the CVNP, NPS, OCC, and Cuyahoga County Planning Commission worked together and completed a study to develop a park along the canal towpath through industrial areas, and connect south to the national park. Some called it an industrial-grade park; they were amazed to find such natural beauty and wildlife so close to heavy industry. By 1995, this concept was expanding, with the Cleveland Metropark development of the Ohio & Erie Canal Reservation (OECR), in which the Towpath Trail would connect south from Cleveland to the Cuyahoga Valley National Park. Partnership later with the Canalway program would result in the Leonard B. Krieger Canalway Center in Cuyahoga Heights within the reservation, which would celebrate the industrial and commercial history of the northern Cuyahoga Valley.

Another initiative came to fruition by the efforts of the Ohio & Erie Canal Scenic Byway Task Force, under joint leadership of the nonprofits OCC and OECC, which in 1993 proposed the creation of a state road scenic system along the O & E canal. This was recognized in 1996 as a state scenic byway, and in 2000 it would become a national scenic byway with appropriate directional signage and tourist promotion.

On August 3, 1995, Congressman Regula submitted a new corridor bill, H.R. 2186, revised to accommodate previous objections and increase support. Nine other states also had heritage-area proposals pending; these were folded together into an omnibus bill passed by Congress on November 12, 1996, and signed by President Bill Clinton. Nine new heritage areas were established, each under its own management. They included the Augusta Canal National Heritage Area in Georgia, the Hudson River Valley National Heritage Area in New York, the South Carolina National Heritage Corridor, the Essex National Heritage Area in Massachusetts, and the Rivers of Steel National Heritage Area in Pennsylvania. That same year, the Miami & Erie Canal Corridor Association was formed.

The section of H.R. 2186 that discusses the Ohio & Erie Canal Corridor is called the Ohio & Erie Canal National Heritage Corridor Act of 1996. The Act called for the formation of several new entities: the Ohio & Erie Canal National Heritage Corridor Committee (OECNHCC), with twenty-one appointees to approve budgets and plans; and the Ohio & Erie Canal Association (OECA), to manage and fund projects of the two existing grassroots nonprofits OCC and OECC. By June 2000, the OECA had completed a management plan and extended the heritage corridor to 110 miles, from Cleveland in the north to New Philadelphia in the south, in three stages:

1) Establish Identity (2000–2006),
2) Pilot Projects (2007–2012), and
3) The Long Term (post-2012).

During the first stage, because the title was too long and people did not understand the concept of the term "corridor," the regional name was changed to The Ohio & Erie National Heritage Canalway. This was often shortened to the Ohio & Erie Canalway (OEC).

In addition to community economic development, OECA grants included funds for building railside shelters for the Cuyahoga Valley Scenic Railroad, serving Akron and Canton, and planned possible extensions southward and northward to downtown Cleveland. Grants have supported preservation programs for historic structures, trails, and towpath construction all along the Canalway. By 1999, OECA grants and Stark County helped connect the Towpath Trail through the most difficult section of Massillon where the towpath no longer existed, by rerouting it along the Tuscarawas River flood levee. When completed, this could complete the Towpath Trail fully through the twenty-three miles of Stark County. In 2003, the trail would be renamed the Congressman Ralph Regula Towpath, in honor of the man who made it possible.

In 1997, in the small canal town of Clinton, in southern Summit County, Tressie McIntosh, founder of the Clinton Historical Society, formed the Clinton Canal Corridor Committee (CCCC) to develop a plan for the two miles of the canal within village limits, which included a guard lock and two lift locks. CCCC partnered with the University of Akron's Interdisciplinary Anthropology Program, which from 1997 through 2001 conducted extensive archeological canal research. CCCC also partnered with Summit County Metro Parks to develop its portion of the canal Towpath Trail, and by 2000, the trail was connected to the Stark County trail in the south. By 2008, eighty-five miles of the Ohio & Erie Towpath Trail were restored and operational. In 2009, a floating towpath was constructed across part of Summit Lake to fill a gap in the trail for hikers and bikers, and it replicated a similar floating towpath originally built there in 1827–28 to fill the same gap for mules.

Many of these local efforts still continue. In 2006, the City of Canal Fulton constructed a visitors and information center, the Canal Fulton Canalway Center, next to the *St. Helena III* boat dock. In 2008, the Stark County Park District opened a new exhibit hall, named the Congressman Ralph Regula Canalway Center. Located in the Exploration Gateway facility at Sippo Lake Park in Canton, Ohio, honoring the man whose vision and support made the redevelopment of the canal possible, it is equipped with high-tech, computerized learning stations about the history of the Ohio & Erie Canal. Ohio has accomplished much in the last fifty years to restore and protect historical areas, with outstanding collaboration between private community groups and town, county, state, and federal organizations.

Since 1996, the Miami & Erie Canal Corridor Association (MECCA) and the Ohio Department of Natural Resources have been working in western Ohio to develop the natural, historical, and recreational resources between Delphos and Piqua as the Miami & Erie Canal Heritage Corridor in a manner similar to the Ohio & Erie Canalway. This fifty-mile section of the canal includes several replica canal boats, many restored locks, and an aqueduct, all accessible by State Route 66, designated as the Miami & Erie Canal Scenic Byway.

These developments are in Ohio. More significantly, similar efforts across the country multiplied this impact on our national culture and heritage exponentially. Thousands of individuals have devoted their lives to this cause, tens of thousands have contributed time and talent, and millions have benefited from these restored recreational and historic locations. Today there are forty-nine National Heritage Areas, and like the Ohio & Erie National Heritage Canalway, a number of them are centered on the restoration and utilization of old canals, including the Illinois & Michigan Canal National Heritage Corridor (97 miles, established 1984), the Blackstone River Valley National Heritage Corridor (45 miles, established 1986), the Delaware & Lehigh National Heritage Corridor (165 miles, established 1988), the Augusta Canal National Heritage Area (8.5 miles, established 1996), and the Erie Canalway National Heritage Corridor (525 miles, established 2000). These are not national parks, but are operated by state governments, nonprofits, or private organizations. However, the National Park Service in many cases provided advice and guidance, and in some areas the NPS logo is used. When added to the Chesapeake & Ohio National Historical Park, these historical areas alone amount to the preservation of nearly 1,100 miles of original U.S. canals, nearly one quarter of the total mileage built in the nineteenth century. Not a bad survival rate after 185 years!

In many instances, this transformation throughout the country was the result of the merger of two powerful civic motivations: pride of historic place and commitment to clean up neglected areas. The abandoned canals often were in the industrial sections of cities and towns originally spawned by the canals. These areas became unfit for residential or parklands and were allowed to deteriorate or become lost. Civic-minded people began to see a double benefit of restoration: they could reclaim their nineteenth-century heritage as well as clean up the environmental mess it left behind. In part, doing so, restored damage done, and in part, it created new lands of utilitarian and economic growth. The communities thus gained a sense of pride and accomplishment.

Those volunteers who worked alongside Ralph Regula in the 1960s can be proud that the *St. Helena II,* Canal Fulton, Stark County, and Ohio were among the earliest pioneers in this great movement to preserve and protect our national canal heritage, including canals that had nearly disappeared into oblivion and which are now populated with operational, replica canal boats to dramatize how they were used. It is also reassuring to know that nineteenth-century canals, which united disparate frontier regions and states into a powerful united nation, shall not perish from the earth but will remain to remind our children and grandchildren of their enormous contribution to America's industrial development and to our high standard of living.

Perhaps most inspirationally, the national preservation, renovation, and re-creation of canal lands and artifacts demonstrate how momentous achievements can be accomplished by dedicated individuals and communities who have no great cause other than to make their own community a little better. As demonstrated by Canal Fulton's *St. Helena II* example, such efforts can, in fact, launch great places that endure forever. How about you? Is there a real canal near you that needs some tender loving care?

Epilogue

On a mostly sunny Saturday in Stark County, Ohio, on August 21, 2010, the City of Canal Fulton and the Canal Fulton Heritage Society hosted the *St. Helena II* Awards and Fortieth Anniversary Celebration in St. Helena Heritage Park. The awards ceremony took place on the deck of the beautifully restored *St. Helena II,* now owned by the City of Canal Fulton and resting proudly beside the Canal Museum, along the Ohio & Erie Canal.

Master of ceremonies was John Harriman, the son of Edward W. Harriman, who had served as master of ceremonies for the dedication and christening of the *St. Helena II* forty years earlier. Keynote speaker was retired U.S. congressman Ralph S. Regula, who in the 1960s inspired Stark County to restore its portion of the canal. He was followed to the podium by Al Simpson, the 1960s author of a weekly column in the *Canton Repository,* "Along the Towpath," who challenged Stark County communities to build an authentic canal boat on the restored canal, a dare accepted by Stark County volunteers who built the *St. Helena II,* in Canal Fulton.

As a current director of the American Canal Society, I then presented a Canal Society Certificate of Authenticity plaque to Linda Zahirsky, president of the Canal Fulton City Council, which operates the current *St. Helena III* canal boat. The certificate reads:

The restored *St. Helena II* at the anniversary celebration in Canal Fulton, August 2010.

> The American Canal Society certifies that this canal boat, *St. Helena II* of Canal Fulton, has met our design criteria as a reasonably authentic historical replica of a canal boat, or type of canal boat, that operated in this vicinity in the nineteenth century.

Jim Guest then presented a cast plaque honoring the significant original builders of the *St. Helena II:* Paul Baird (1896–1986), Clyde Gainey (1895–1987), Carroll Gantz (b. 1931), Jim Guest (b. 1935), Gale Hartel (1924–1983), Ed Harriman (1921–1981), Dick Mohler (b. 1927), Pete Neidert (1903–1977), Roy Preece Sr. (1918–1982), and Walter Smith (1902–1982); and significant contributors Stark County Historical Society, Louisville Sportsman's Club, Hoover Foundation, and Timken Foundation.

The president of the Canal Fulton Heritage Society, John Hatfield, then presented a similar plaque honoring the *St. Helena II* Restoration Committee members John Hatfield, Todd Gibbs, Carolyn Swinehart, Ina Hatfield, Dan Mayberry, Libby Wilson, Dick Mohler, Jim Guest, Greg Mayberry, Larry Turner, and Tom Walters; and significant contributors Clayton Hopper, Rosemary Benson, Dennis Mayberry, Rosemary Gibbs, Pat Pentello, Bill Loretto, Ken Bailey, Linda Shindler, Jeff Boak, Linsey Concrete, CTI Environmental, Ohio Edison, the Canal Society of Ohio, and Stark County Parks. This group saved the *St. Helena II,* which was in total ruin in 1988, from certain destruction by rebuilding her from the ground up, starting in 1998.

All three plaques were to be mounted on the restored *St. Helena II.* The occasion provided a special reunion for original contributors and their families, who were treated to a ride on the *St. Helena III* and a picnic luncheon in the park. The event was also a celebration for the Canal Fulton community, which has successfully maintained and expanded its historical canal boat legacy, started forty years before.

APPENDIX A

"Canal Nostalgia," by James Dillow Robinson (1899-1980)

This text is from a copy of the poem Robinson distributed to attendees on the July 11, 1970, *St. Helena II* celebration.

In the days when red men trod the trails
and on the plains Honest Abe was splitting rails,
In the days when covered wagons headed west,
In the Ohio wilderness the old canal was at her best.
With quarried stone the pioneers had built the locks
And hand-hewn timbers made the aqueducts.
By hand they dug the channel and formed the towpath.
They paid the cost, braved the fever, and finished the task.
In 1828, through this Tuscarawas valley green,
First flowed the old canal, a slow and lazy stream.
Through this valley the good boat St. Helena made her way—
She was the pride and joy of Tuscarawas folks that day.
The old St. Helena—she was class and she was style—
And her missions were important and worthwhile.
Aboard, the Captain greeted each new-born day with smiles,
And journeyed down the Tuscarawas valley for many pleasant miles.
Now the Buckeye State was on the move and she would grow
And the history of the old Canal has proven this was so.

She served well her state from that early age
And is deserving of her honors on Ohio's history page.
The pioneers had labored hard this old canal to build
And with the good St. Helena their hearts were thrilled.
For now they knew their dream would soon come true
And boats would journey from the Erie lake to the Ohio blue.
But today like an old shoe worn and cast away,
For the most part the old canal is lost and in decay.
Only remnants remain to tell us of the frontier days
When canal boats and their cargo passed this way.
Traces of the old towpath, a few moss-covered locks,
Raceways, spillways, and tumbled down aqueducts,
Remind us that here once flowed that pleasant stream—
The old Ohio Canal of which pioneers dared to dream.
Some old-fashioned clocks no longer tell the time or tick away,
But Canal remnants tell the story and the times of bygone days
When in the Ohio wilderness the old Canal played her part
Built by determined pioneers, strong and brave of heart.
Through the valleys, o'er the summits, beside the winding rivers,
Flowed the old Canal—a gentle, pleasant stream,
Bringing life and more abundance like the constant waters of an everflowing spring.
Came the rails of progress and the canal "gave up the ghost"
And her calm and peaceful waters had no further need of boats.
Other plans of men had ended the "lake to river" run,
No more would boats go through the locks; the old Canal was done.
While others traveled onward on wagons to the west,
The old Canallers stayed to build and open Ohio wilderness.

Today, in memory of Pioneers, Settlers, Canallers all,—
We tell their proud story and preserve here the Ohio Canal.

J. DILLOW ROBINSON
Independence, Ohio

This dedication celebration of the canal boat, *St. Helena II,* at Canal Fulton, Ohio, revives the appreciation of the work of early settlers and pioneers of Ohio, and keeps alive their deeds for future generations.

J. DILLOW ROBINSON

APPENDIX B

Honorable Ralph Regula's Dedication Address, July 11, 1970

What a great day! I am an eternal optimist but even I began wondering if this would ever get done.

July 4, 1827, 143 years ago almost to the day—the canal boat "State of Ohio" leaves Akron for Cleveland on the new Ohio-Erie Canal. I am sure for that crowd it must have held all the thrill of a modern-day space blastoff.

Elsewhere in Ohio forests still ring with Indian whoops. A new era dawns—soon a scar will cut across the face of the land from Cleveland to Portsmouth—309 miles long—a big ditch that will transform a territorial wilderness into the third most populous state of that age in but one generation. An era that will bring a flood of immigrants to settle the richest lands of Ohio. A turnpike of water that would connect the rich Ohio farmland, mixed resources and magnificent forests with the world's markets through the Great Lakes and the Ohio-Mississippi Valley.

Ohio's industrial backbone was to be built upon the skeleton of the thousand miles of Ohio canals. It becomes a lifeline for the state. Villages become cities—prosperity and progress become synonymous with the Buckeye state. The roll call of cities—Cleveland, Akron, Massillon, Navarre, Dover, New Philadelphia, Coshocton, Columbus, Chillicothe, and Portsmouth. In the west—Toledo, Defiance, Dayton, Middletown, Hamilton, and Cincinnati.

Yesterday's turnpike is born—Ohio has a window to the world. A magnificent engineering feat—that should rank as one of the wonders of the world of that age. A triumph to the determination of individuals—a testimonial to the courage, resourcefulness, and boundless energy of the men who opened up the Northwest Territory. Men such as:

- George Washington—who first proposed an inland hydraulic highway to connect the Mississippi and the St. Lawrence waterways.
- DeWitt Clinton—whose Erie Canal inspired Ohioans.
- Ohio governor Ethan Allen Brown—"The Father of Ohio Canals" who in 1818 in his inaugural address emphasized the need for a cheaper way to market produce.
- Nathan Gulford—Caleb Atwater—who as legislators were to engineer a coalition that would provide a bill authorizing a commission to investigate a canal route for Ohio and at the same time form another commission to devise a system of public education—a compromise that would have lasting benefits.
- Alfred Kelley—Lawyer, legislator, banker, engineer and statesman—who would pledge his personal fortune to guarantee the bonds essential to building a canal system.

And we cannot forget the thousands of rugged individuals whose labors and, in some cases, lives, built a muddy ditch that brought the world to the wilderness.

July 11, 1970, Canal Fulton, Ohio—a village whose historic registers would record the names of John Quincy Adams, William Henry Harrison and William McKinley—all canal travelers. The beginning of a new golden era—a renaissance of canal interest has begun. A window to the past replaces a window to the world. The call of the canawlers will once more ring through the historic Tuscarawas Valley. Ironically this canal that transformed a wilderness to an industrial state would now provide a ribbon of wilderness for all ages.

Today we launch *St. Helena II,* the only completed authentic canal boat in the nation—a truly historic occasion for us and for Ohio. We add a jewel to Stark County's crown—one that can proudly take its place with the Stark Wilderness Center, the Stark Historical Society and the Canton Hall of Fame. Today we add a chapter to the Stark County Story.

How has the "Couldn't Be Done Dream" become a "Could Be Done Reality?" Again, as in 1827, it is a triumph to INDIVIDUAL DETERMINATION. The key words being INDIVIDUAL and DETERMINATION.

In a recent issue of *Newsweek* analyzing the American crisis, historian Andrew Hacker states, "We have in short become a loose aggregation of private persons who give higher priority to our personal pleasures than to collective endeavors. Americans no longer display the spirit that transforms a people into a citizenry and turns territory into a nation.

The success of this project belies his pessimism and proves that the unselfish spirit that made America great still exists here in Stark County. How many have shared in this spirit? How many have made this day possible? Perhaps hundreds of individuals that will never be formally recognized. And yet we have recorded over five hundred names of individuals and corporations whose contributions have provided this historic first. Among these are:

- A newspaperman, a conservationist, a modern-day ecologist with a gifted pen that captured the imagination of a multitude and inspired disciples to the cause throughout Ohio.
- A Tuscarawas County canawler whose exciting tales of canal days enriched the dreams of a canal boat and who now himself is part of the canal legend.

- A retired county commissioner with the vision to preserve for the future what he knew only as a memory.
- An engineer, a curator, a cartographer of the canals and most importantly a believer that yesterday's dream could become today's reality.
- A businessman turned tree hunter and scavenger.
- A community's elder statesmen turned blacksmiths and masters of many trades.
- A twentieth century designer who is the reincarnation of a nineteenth century canal boat builder—extraordinary.
- Louisville Sportsmen turned canalboat builders.
- A museum director with a low resistance to canal fever and a genius for organization.
- Corporations, organizations and foundations too numerous to mention who put their labor, their materials, their money and most importantly their faith in the will of individuals.
- The excellent cooperation of state and county officials cleared many hurdles.
- The hard-hat cranemen with the soft touch whose skills carefully lowered the St. Helena II to its watery christening while a crowd of onlookers watched breathlessly.
- A patron of this community, a great lady in any age, whose generosity sparked tomorrow's dream of a transportation museum at Lock 4 Park.
- And most importantly—a community blessed with enthusiasm and leadership.

Today we launch more than a canal boat. The *St. Helena II* is a symbol that the pioneer spirit of individuals can achieve the impossible—and that it can be achieved using the resources and energies of the community alone. The thing is, we stuck together and we finished it.

The success of this project stands as a challenge to all of us—to continue to preserve the heritage of our past and to challenge us to great goals in the future. Borrowing a phrase from one of Stark County's well-known industries we launch a symbol of our proud past and look forward to an exciting future that will include continued enhancement of Stark's ribbon of wilderness, its twenty-three miles of canal lands.

APPENDIX C

Transcript of recorded lecture played on the *St. Helena II* on July 11, 1970

The Stark County Historical Society welcomes you aboard the *St. Helena II* of Canal Fulton. In the next forty-five minutes, we would like to take you 2½ miles and 140 years back into history, when the Ohio canals opened the state to commerce, industry, and settlers. Your next forty-five minutes will be spent on this original portion of the Ohio-Erie Canal, which was begun on July 4, 1825. When completed, it stretched 309 miles from Cleveland on Lake Erie to Portsmouth on the Ohio River, passing through Peninsula, Akron, Canal Fulton, Massillon, Bolivar, Dover (then Canal Dover), Roscoe, Dresden, Newark, Lockbourne, Circleville, Chillicothe, and Waverly—to name just a few towns. It connected to the Pennsylvania & Ohio Canal at Akron, which went to New Castle via Kent, Ravenna, Warren, and Youngstown; with the Sandy & Beaver Canal at Bolivar, which went to Marietta via Zanesville and McConnelsville; with the Walhonding Canal at Roscoe leading to Brinkhaven; with the Hocking Valley Canal at Carroll, leading to Athens via Lancaster, Logan, Nelsonville; and to Columbus by feeder canal.

It is difficult to comprehend the magnitude of the task of building the Ohio canals. To traverse the state across the summit near Akron, a height of 390 feet above Lake Erie, 152 locks were required. Lock 4, which you are about to visit, is the fourth lock south of the summit. The canal route had to be carefully selected to provide a constant water supply, and the canal beds excavated to an average four-foot depth; twenty-six feet wide at the bottom, and forty feet wide at the surface. The approximate cost of building the canal was $4½ million.

Canal boats used were generally of two types; passenger boats, called "packets," and freight barges. Packets, patronized by the wealthy, could travel the 309-mile distance from Cleveland to Portsmouth in eighty hours, at a cost of about 4 cents per mile. They had a single cabin enclosure the length of the boat, and accommodated about one hundred people. Freight barges were also used for less wealthy passengers. In addition, they carried fifty to eighty tons of cargo, such as lumber, stone, wheat, or coal. The fare and proposed speed advertised to passengers was "a cent and a half a mile, and a mile and a half an hour."

St. Helena II is authentically designed to represent a typical freight barge. She is sixty feet on the keel and thirteen feet wide, weighing twenty tons empty and drawing eighteen inches of water. Her framing timbers, keel, hull, and decking are all solid white oak. Her aft cabin originally was used as the traveling home of the captain and his family. There you will find a cook stove, serving counter, generous bunk, and cupboard space. The stable cabin where you boarded was originally used for mules. There you can see feed bins and a grain storage compartment. The forward cabin was intended for crew or passengers. It contains a bunk and potbelly stove. The decks where you are seated were originally cargo holds, and were usually piled high with wheat, lumber, or stone. The square wooden structures near the cabins are pump wells, used to bail out excess water from the bilge with a pump.

Because of the cargo, the crew and passengers had to walk the overhead catwalks to enter the cabins. Overhead hatches, and ladders in the fore and aft cabins show how this was done. The capstan, located on top of the center cabin, is used to raise and lower the gangplank. This, as well as all fittings on the boat, was handmade by volunteers in a makeshift blacksmith shop.

The captain's position is on the tiller deck at the stern. The crew is on the fore-deck at the bow to spot danger and to help turn the boat with pike poles. Two mules on the towpath are hitched tandem to the 150-foot tow rope, which is attached to a fitting on the forward cabin called a deadeye. The mule driver on shore controls the mules. Mules were chosen because of their high endurance and low maintenance. Two to three were used, depending on the weight of the cargo. Speed was limited to four miles per hour to prevent erosion of the banks. Navigation rules required that packets had precedence over cargo boats. Also, when boats met, the downstream boat stopped and let its tow rope sink to the bottom, giving the upstream boat the right of way. Boats generally cost from $2,000 to $4,000 each, including mules ($50,000 to $100,000 in 2010 dollars).

St. Helena II was built over a 3½-year period. Starting in the winter of 1967, oak trees were felled to provide timbers, and the keel was laid in 1967 on the bank across from the Park dock. By spring of 1968, the framing was completed and hull planking was begun. The 1½ -inch thick oak hull planks were boiled, bent, and spiked to the frames. During the winter of 1968–1969, planking was completed under a plastic shelter, and the seams were caulked with oakum and cotton. July 1969 saw the hull lifted and rolled into the canal for splashdown. Work on the decks and cabins continued through the winter of 1969–1970, and by June of 1970, St. *Helena II,* the first authentically built canal boat in the country, was ready for her shakedown cruise. It is estimated that over fifteen thousand hours of volunteer labor, and a thousand hours of paid labor were required. Nearly $25,000 was received in donations from a long list of donors, the largest being the Stark County, Timken, and Hoover Foundations.

Future plans include operating the boat through Lock 4. To do this, a wooden towpath needs to be constructed around the large willow trees for the mules. Also planned is a restoration of the McLaughlin drydock, just south of the Community Park, where the boat can be serviced and maintained properly. The Stark County Historical Society hopes to build a museum at Lock 4, where the boat may be viewed year-round. It is not inconceivable that *St. Helena II* could operate into northern Massillon, with the restoration of the towpath and revisions to a few culverts and bridges. Or, north another mile through Canal Fulton to the feeder canal. A total of eight miles of canal awaits *St. Helena II;* and perhaps her sister ship!

When you arrive at Lock 4, you will notice the two sets of lock gates. The lower gates are closed to hold the water in the lock; the upper gates open to receive the boat. When the boat is within the eighty-foot lock walls, the upper gates are closed, and the butterfly valve in the lower gates open to slowly allow the water level in the lock to sink ten feet. When this is achieved, the lower gates open to permit the boat to pass through. We hope you enjoy your short journey on the Ohio-Erie Canal. Please take advantage of the cabin windows to view the interior of the cabins. The crew members or hostesses will be happy to answer any questions you might have. We hope you will return again to see the canal in a variety of seasonal moods. Thank you.

Notes

2. OHIO CANALS

1. Ohio Canal Act of 1825, quoted by Jack Gieck, *A Photo Album of Ohio's Canal Era, 1825–1913* (Kent: Kent State University Press, 1988), 3.

2. George Washington, in a letter to Governor William Livingston of New Jersey, December 1776, quoted by David Hackett Fischer in *Washington's Crossing* (New York: Oxford Univ. Press, 2004), 134.

3. DECLINE AND SURVIVAL

1. Bill Yoder, "History or Gossip? A Bit of Whimsey," *Canal Currents* (Autumn 1976): 11.

4. STARK COUNTY, OHIO

1. Ralph Regula, quoted by Al Simpson, "Along the Towpath," *Canton Repository*, May 10, 1964, 4, 10 (hereafter, these articles are cited as "Along the Towpath").

2. Ibid., 4

3. Simpson, "Along the Towpath," Nov. 22, 1964, 10.

4. Hahn, *Canal Terminology of the United States*, 107.

5. Simpson, "Along the Towpath," Apr. 4, 1965, 10.

6. Terry Woods and Todd A. Clark, statements to author, Apr. 2011.

7. Simpson "Along the Towpath," June 13, 1965, 39

8. Simpson, "Along the Towpath," May 1, 1966, 95

6. CONSTRUCTION BEGINS

1. Ed Harriman, in the first issue of the *Canawler* (Summer 1967): 4, announcing, as Canal Fulton Exchange Bank president, the purpose of the bank's sponsorship of the publication.

2. Simpson, "Along the Towpath," June 30, 1968, 201.

3. "Canal 'Craze' Developing in Ohio," *Canton Repository,* Nov. 17, 1968.

8. DEDICATION

1. James Dillon Robinson, "Canal Nostalgia," from a flier distributed by its author on the dedication of the *St. Helena II,* Canal Fulton, Ohio, Saturday, July 11, 1970. (See appendix A.)

2. Newspaper articles Richard Sicha provided to Peg Bobel and thence to author, and information from Robinson interviews with Terry Woods.

3. Ohio state senator Ralph Regula, as quoted by John C. Harriman in his book, *Sticking Together to Build the Canal Boat* St. Helena II (Milford, Ohio, Gordon Bernard Company, LLC 2007–10), 27, 30-31. (See appendix B.)

4. Transcript of recording played on the *St. Helena II*'s first public ride, July 11, 1970. (See appendix C.)

10. OHIO & ERIE CANALWAY

1. Ohio state senator Ralph Regula, as quoted by John C. Harriman in his book, *Sticking Together* (Milford, Ohio: Gordon Bernard Company, LLC, Milford, Ohio, 2010), 31.

2. Lynn Metzger and Peg Bobel, eds., *Canal Fever: The Ohio & Erie Canal, from Waterway to Canalway* (Kent, Ohio: Kent State University Press, 2009), 258–59. Throughout this chapter, information about canal restoration is taken from Metzger and Bobel's book, chapters 9, 14, 16, and 17.

Bibliography

American Canal Society. *American Canals*, 1972–2010.

Bunch, Bryan, and Alexander Hellemans. *The Timetables of Technology*. New York: Simon & Schuster, 1993.

Burk, Kathleen. *Old World, New World*. New York: Atlantic Monthly Press, 2008.

Erie Canal Sesquicentennial Committee of the City of Rome, *40′ x 28′ x 4′: The Erie Canal—150 Years*. Rome, N.Y.: Oneida County Erie Canal Commemorations Commission, 1967.

Fischer, David Hackett. *Washington's Crossing*. New York: Oxford University Press, 2004.

Frey, Russell. *Rogues' Hollow History and Legends*. Rittman, Ohio: Rittman Press, 1958.

Gantz, Carroll. *St. Helena II: The Design and Construction of the First Authentically Reconstructed Canal Boat in the U.S. 1967–1970*. Johns Island, S.C.: self-published, 2004.

———. "St Helena II—Her Design, Construction, Operations, Demise, Restoration and Replacement by St. Helena III." PowerPoint presentation, March 2008.

Gieck, Jack. *A Photo Album of Ohio's Canal Era, 1825–1913*. Kent, Ohio: Kent State University Press, 1988.

———. *Early Akron's Industrial Valley: A History of the Cascade Locks*. Kent, Ohio: Kent State University Press, 2008.

Hahn, Thomas. *The C & O Canal: An Illustrated History*. York, Penn.: American Canal and Transportation Center, 1981.

Hahn, Thomas Swiftwater, and Emory L. Kemp. *Canal Terminology of the United States.* Morgantown, W.V.: Institute for the History of Technology and Industrial Archaeology, Eberly College of Arts & Sciences at West Virginia University, in cooperation with the National Park Service and West Virginia University Press, 1999.

Harriman, John C. *Sticking Together to Build the Canal Boat* St. Helena II. Milford, Ohio: Gordon Bernard Company, 200–210.

———. "Building the *St. Helena II*—Canal Boat Construction through the Photography of Ed Harriman, Gale Hartel, and Clyde Gainey." PowerPoint presentation, 2007.

Kurlansky, Mark. *Salt: A World History.* New York: Penguin, 2003.

Metzger, Lynn, and Peg Bobel, eds. *Canal Fever: The Ohio & Erie Canal, from Waterway to Canalway.* Kent, Ohio: Kent State University Press, 2009.

Pennsylvania Canal Society. *Canal Currents,* 1968–1980.

Shank, William H. *Three Hundred Years with the Pennsylvania Traveler.* York, Penn.: American Canal and Transportation Center, 1976.

Simpson, Al. *Along the Towpath.* Akron, Ohio: University of Akron Libraries, 2003.

Van Alstyne, Henry A. *Canals and Navigable Rivers of the United States and Canada.* Albany, N.Y.: Brandow Printing Company, state printers, 1905.

Wakefield, Manville B. *Coal Boats to Tidewater: The Story of the Delaware and Hudson Canal.* Grahamsville, N.Y.: Wakefair Press, 1971.

Wilcox, Frank, and William A. McGill. *The Ohio Canals.* Kent, Ohio: Kent State University Press, 1969.

Woods, Terry. *Ohio's Grand Canal: A Brief History of the Ohio & Erie Canal.* Kent, Ohio: Kent State University Press, 2008.

———. *The Ohio & Erie Canal: A Glossary of Terms.* Kent, Ohio: Kent State University Press, 1995.

Zimmerman, Albright G., *Pennsylvania's Delaware Division Canal—Sixty Miles of Euphoria and Frustration,* Easton, Penn.: Canal History and Technology Press, 2002.

Illustration Credits

Canal Fulton Heritage Society: 102; *Canton Repository:* 101; Edward W. Harriman Family Collection, courtesy of John C. Harriman: 45, 51, 61, 67, 98, 101; McLaughlin Family Collection: 54, 56; Ohio Historical Society: 15, 34; Victoria Neidert-Hammer: 38.

Note: all illustrations not otherwise credited are from the author's collection or are in public domain.

Index